PLANT SCIENCE RESEARCH AND PRACTICES

# *LACTUCA*

# CULTIVATION AND USES

# Plant Science Research and Practices

Additional books and e-books in this series can be found on Nova's website under the Series tab.

PLANT SCIENCE RESEARCH AND PRACTICES

# *LACTUCA*

# CULTIVATION AND USES

JAN KRÜGER
EDITOR

## NOTICE TO THE READER



**Library of Congress Cataloging-in-Publication Data**

Library of Congress Control Number: 2020934950
ISBN: 978-1-53617-729-9

*Published by Nova Science Publishers, Inc. † New York*

# CONTENTS

# PREFACE

The *Lactuca* genus, which is characterized by a variety of forms including annual, biennial and perennial, glabrous or pubescent herbs with abundant latex, belongs to the Asteraceae family. *Lactuca* is an essential source of phytochemicals and is used in traditional medicine.

The authors discuss *Lactuca* sativa's anti-inflammatory, antioxidant, antidiabetic, antimicrobial, analgesic, sedative and neuroprotective effects.

The antibacterial effects of *Lactuca* sativa extracts are evaluated using the microwell dilution method against a wide range of clinical isolated bacteria, and the antioxidant effects are tested using ferric-reducing antioxidant power and catalase activity assay.

The genetic transformation of lettuce is examined in conjunction with the applications of this transformation in plant biotechnology.

The potential for Ulva *Lactuca* to be used as a bioindicator of contamination is studied, as well as its capability to be employed for the removal of pollutants.

The authors explore the hypothesis that if a sufficient amount of free vitamin B12 can be incorporated into lettuce leaves grown under hydroponic conditions, lettuce could become an excellent source of vitamin B12 for vegetarians and elderly people.

Chapter 1 - The *Lactuca* genus, which is characterized by a variety of forms including annual, biennial and perennial, glabrous or pubescent herbs with abundant latex, belongs to the Asteraceae family. Overall taxonomic literature has shown that 100-111 species represent the genus. The members are found primarily in warm and temperate regions mostly in the Northern Hemisphere. Among the all known species in the group of leafy vegetables, *Lactucasativa* is considered as the most important vegetable. In many countries around the world, lettuce is cultivated commercially and is usually grown as a vegetable. The species is known for its high genetic diversity due to its polyphyletic origin and a complex cycle of domestication. Lettuce is the United States ' second most valuable crop and is a rich source of natural ingredients including ascorbic acid, vitamin A, vitamin K, β-carotene, lutein, and zeaxanthin. Besides, it is also regarded as a source of many minerals such as calcium, phosphorous, iron and copper. The *Lactuca* species are also essential sources of different phytochemicals and are used in traditional medicines for curing many ailments including inflammation, pain, stomach problems including indigestion and lack of appetite, bronchitis and urinary tract infections. The species show various pharmacological activities including antioxidant activity, anti-cancer activity, and anti-microbial activity and are also used for the treatment of neurodegenerative disorders. *Lactuca*rium, a white milky compound isolated from wild lettuce (*L. virosa*), delivers similar effects as opium and is often referred to as "Opium lettuce" due to its pain-relieving activity. Aside from pain, the species is recommended as a natural treatment for a variety of disorders such as respiratory conditions, menstrual cramps, arthritis, cancer, insomnia, poor circulation, and urinary infections. Today there are number of wild lettuce products available in the market which contain extracts of plants leaves, seeds and milky sap and are considered beneficial for various human health disorders including anxiety, breathing issues, poor sleep, and joint pain.

Chapter 2 - *Lactuca sativa,* commonly known as lettuce, belongs to the Asteraceae family. It is used as both delicious vegetable and important folk medicine. It is low in carbohydrate and fat contents with high water contents. Lettuce is rich with minerals and vitamins including vitamin A, C, E, iron, potassium, calcium and phosphorus. It also contains selenium which has medicinal properties for the prevention of colon, prostate and lung cancers. This chapter emphasizes the medicinal properties of lettuce and its active components for different biological actions, for example, anti-inflammatory, antioxidant, antidiabetic, antimicrobial, analgesic, sedative and neuroprotective effects. The therapeutic potential of this plant is primarily due to sesquiterpene lactones which are considered as the most active compounds of lettuce. These compounds impart health benefits by providing a balanced diet and by acting as therapeutic agents, due to their strong curative potential against heart and cancer diseases. Lettuce is also a rich source of phytochemicals and plant secondary metabolites. These phytochemicals in lettuce have been reported for analgesic, hypnotic, sedative and anticonvulsant properties. Dietary phytochemicals are of specific interest due to their significance as antioxidants. Later in the chapter, the authors have discussed different techniques that have been applied to identify phenolic and phytochemicals compounds in lettuce like liquid chromatography and mass spectrometry. A detailed primary and secondary metabolite profiling of lettuce has also been described. Because lettuce has strong medicinal potential, its utilization in dietary supplementation can be further investigated. Moreover, clinical studies are also needed to assess the efficiency and safety of lettuce preparations and its isolated compounds for the therapeutic application.

Chapter 3 - *Lactuca sativa* var crispa has been more present in the nutrition in the last few years. This vegetable has a high prophylactic effect due to high level of biologically active matters in it. Its strong attractive colors indicate the presence of phenol compounds and therefore good antiradical activity. Antibacterial, anticoagulant and antioxidant activities of *Lactuca sativa* var. crispa extracts were investigated. The antibacterial activity was evaluated using microwell dilution method against a wide range of clinical isolated bacteria. Antioxidant activity of *Lactuca sativa* extracts were tested using Ferric-Reducing Antioxidant Power (FRAP) and Catalase Activity Assay (CAT). Among tested extracts, aerial part extracts showed the best antibacterial activity with Minimal Inhibitory Concentration (MIC) ranged from 0.062 to 1.5mg/ml. In addition, such extracts exhibited the highest anticoagulant and antioxidant activities. The results provided an evidence that the studied vegetable might, indeed, be potential sources of natural antioxidant, anticoagulant and antimicrobial agents.

Chapter 4 - Lettuce (*Lactuca sativa* L.) is a worldwide vital leafy crop that is cultivated globally. As a result of the speedy progression of human population and their desire to grow, scientists have conducted many studies especially on the genetic transformation or genetic engineering of this plant. The main focus of the current lettuce cultivation programs is to improve horticultural as well as pharmacological characteristics of the plant such as enhancing quality, protection to early bolting, produce disease-free plants and production of important proteins of biopharming. These characteristics can be improved by genetic engineering as well as by adopting different transgenic approaches. The development of efficient and liable tissue culture methods is of prime importance for the successful accomplishment of the genetic transformation of the lettuce. Advances in micropropagation system and transformation methods have aided in increasing the transformation efficacy and constant expression of transgenes in lettuce. Although there is a lot of research that has been done on lettuce with the main purpose to produce pharmaceutically important proteins and vaccines, detailed comprehensive reviews on the basic transformation strategies and major regeneration system of lettuce plant are still missing. Moreover, chloroplast transformation offering unique benefits including high-level expression of foreign protein also needs to be considered. This chapter is focused on the genetic transformation of the lettuce including both nuclear as well as chloroplast transformation system along with major regeneration system and applications in plant biotechnology.

Chapter 5 - Pollution of natural waters has become a major issue all over the world. As a result, scientists are studying new and alternative technologies not only to identify the presence of different pollutants in the environments but to remove them from waters and industrial effluents. In these senses, in recent years various red, green and brown seaweeds were investigated as potential bioindicators of contaminants in the environment; as well as their use in remediation process. The green marine macroalga, *Ulva lactuca* (Chlorophyta) is very common at coastal areas, and is a bloom forming macroalgae that occur in shallow estuaries. This specie has a high bonding affinity with heavy metals and several organic compounds due to the presence of different functional groups on its cell walls (such as carboxyl, hydroxyl, phosphate or amine) that can bind metal and/or organic ions. These characteristics makes it suitable to be used as a bioindicator of marine pollution and for remediation purposes. In general, the investigations are focused on the application of dried algae biomass to promote the removal of different contaminants, mainly metals from aqueous solutions (biosorption); however, the use of living organisms (bioaccumulation), may be advantageous. While growing, living macroalgae can remove simultaneously pollutants and nutrients (nitrates and phosphates) from wastewaters and capture $CO_2$ emissions. In this context, the present chapter will conduct a comprehensive inquiry on the advances related to the possibility of *Ulva lactuca* to be used as a bioindicator of contamination as well as its capability to be employed for the removal of pollutants from the surrounding environment. It will review these different approaches, and analyse the advantages and disadvantages of each as exposed on the literature, as well as the challenges that these alternatives confront as sustainable remediations technologies.

Chapter 6 - Lettuce (*Lactuca sativa* L.) is a popular leafy vegetable that can be easily cultivated and is a good source of various nutrients. However, plants, including vegetables, generally contain no or trace amounts of vitamin $B_{12}$ as they do not require this vitamin $B_{12}$ for growth. Therefore, if a sufficient amount of free vitamin $B_{12}$ can be incorporated into lettuce leaves grown under hydroponic conditions, it would become an excellent source of free vitamin $B_{12}$ for vegetarians and elderly people. Lettuce leaves were grown for 30 days under hydroponic culture conditions and treated with 5 μM vitamin $B_{12}$ for 24 h, after which the vitamin content in the leaves increased significantly from nondetectable levels to 164.6 ± 74.7 ng/g fresh weight. Thus, the recommended dietary allowance of vitamin $B_{12}$ for adults (2.4 μg/day) in the United States and Japan can be met by consuming only two or three of these fresh leaves. Moreover, the majority (86%) of vitamin $B_{12}$ in these leaves was recovered in the form of free vitamin $B_{12}$ fractions. These results indicate that vitamin

$B_{12}$-enriched lettuce leaves could be used as an excellent source of free vitamin $B_{12}$, particularly in elderly people with food-bound vitamin $B_{12}$ malabsorption.

In: Lactuca: Cultivation and Uses
Editor: Jan Krüger
ISBN: 978-1-53617-729-9


*Chapter 1*

# *LACTUCA* L.: WORLD DISTRIBUTION AND IMPORTANCE

***Mohammad Saleem Wani[1], PhD, Younas Rasheed Tantray[1], PhD, Ishrat Jan[2], PhD, Vijay Kumar Singhal[1] and R.C.Gupta[1]***

[1]Department of Botany, Punjabi University Patiala, India
[2]University of Kashmir, Srinagar, India

## ABSTRACT

The *Lactuca* genus, which is characterized by a variety of forms including annual, biennial and perennial, glabrous or pubescent herbs with abundant latex, belongs to the Asteraceae family. Overall taxonomic literature has shown that 100-111 species represent the genus. The members are found primarily in warm and temperate regions mostly in the Northern Hemisphere. Among the all known species in the group of leafy vegetables, *Lactucasativa* is considered as the most important vegetable. In many countries around the world, lettuce is cultivated commercially and is usually grown as a vegetable. The species is known for its high genetic diversity due to its polyphyletic origin and a complex cycle of domestication. Lettuce is the United States ' second most valuable crop and is a rich source of natural ingredients including ascorbic acid, vitamin A, vitamin K, β-carotene, lutein, and zeaxanthin. Besides, it is also regarded as a source of many minerals such as calcium, phosphorous, iron and copper. The *Lactuca* species are also essential sources of different phytochemicals and are used in traditional medicines for curing many ailments including inflammation, pain, stomach problems including indigestion and lack of appetite, bronchitis and urinary tract infections. The species show various pharmacological activities including antioxidant activity, anti-cancer activity, and anti-microbial activity and are also used for the treatment of neurodegenerative disorders. *Lactuca*rium, a white milky compound isolated from wild lettuce (*L. virosa*), delivers similar effects as opium and is often referred to as "Opium lettuce" due to its pain-relieving activity. Aside from pain, the species is recommended as a natural treatment for a variety of disorders such as respiratory conditions, menstrual cramps, arthritis, cancer, insomnia, poor circulation, and urinary infections. Today there are number of wild lettuce products available in the market which contain extracts of plants leaves, seeds and milky sap and are considered beneficial for various human health disorders including anxiety, breathing issues, poor sleep, and joint pain.

**Keywords:** *Lactuca*, cultivation, distribution, uses

## INTRODUCTION

The genus *Lactuca* is classified in the Asteraceae, subfamily Cichorioideae, tribe Lactuceae Cass., subtribe LactucinaeDumort (Bremer et al., 1994, Judd et al., 2007, Kadereitet al., 2007). The family Asteraceae has a global distribution (Heywood 1978), with about 1535 currently accepted genera and 23,000 species (Judd et al., 1999). Generic limits are often problematic, and several large genera are frequently divided into numerous segregates (Bremer et al., 1994). The family is partitioned into 17 tribes, which are often arranged into 3 subfamilies (Judd et al., 1999). Tribe Lactuceae (=Cichorieae) is identified by ligulate capitula and copious milky latex and is divided into 11 subtribes, represented by 98 genera and more than 1550 species. Subtribe Lactucinae involves 17 genera, including *Lactuca*, and comprises about 100-111 wild species. (Bremer et al., 1994, Stojakowska et al., 2013, The International Plant

Names Index and World Checklist of Selected Plant Families 2020). Frietema (1994) considered the common name of the species belonging to the genus *Lactuca* to be "lettuce." The genus *Lactuca* is characterized by a range of forms, including annual, biennial, and perennial, glabrous or pubescent herbs with abundant latex, rarely shrubs, rhizomatous, sometimes with underground stolons or with fusiform and/or tuberous roots. The generic concept of *Lactuca* was very much expounded by Feráková (1977), at least for European species. However, the complex of *Lactuca* spp. originating on other continents is not very well clear up from a taxonomic viewpoint (Koopmanet al., 1998, Doležalováet al., 2002, Lebedaet al., 2007). A classification of these species, based on taxonomic and geographical concepts, was expounded by Lebeda (1998) and Lebeda and Astley (1999). In this classification, the genus is divided into seven sections and two (African and North American) geographical groups. Recently, the genus concept was discussed in detail by Koopmanet al., (1998).

Despite the lack of studies focused on the entire *Lactuca* genus, there have been many studies focused on cultivated lettuce and closely-related wild species. These studies focused on aspects of interest for lettuce breeding to improve growth related to abiotic and biotic stresses using genetic resources from wild lettuce species (Hartman et al., 2012, 2014, Jeuken et al., 2008, van Treuren et al., 2011). Zohary (1991) set up an idea of the 'lettuce gene pool' and Koopman et al., (1998, 2001) modified Zohary's lettuce gene pool concept and gave the first molecular phylogenetic relationships among *Lactuca* species based on nrDNA ITS-1 and AFLPs. Koopman et al., (1998) portrayed *L. sativa*, *L. serriola L.*, *L. dregeana* DC., *L. aculeate* Boiss. and *L. altaica* Fischer et C.A. Meyer as the primary gene pool, *L. virosa* L. and *L. saligna* L. as the secondary gene pool, and *L. quercina* L., *L. viminea*, *L. sibirica* Benth. Ex Maxim. and *L. tatarica* (L.) C.A. Meyer as the tertiary gene pool. Apart from Koopman et al., (2001) and Wang et al., (2013), there is finite information about the molecular phylogenetic relationships within the genus *Lactuca*, chiefly for the African species since they were first described (Jeffrey 1966, Stebbins 1937).

The genus *Lactuca* is disseminated in temperate and warm regions, for the most part in the Northern Hemisphere (Europe, Asia, Indonesia, North and Central America, Africa). Lebeda et al., (2004) provided an overview of the biogeographical distribution of wild *Lactuca* species based on the accessible literature data and presented that Asia (containing 51 species) and Africa (containing 43 species) are the two centers of diversity for *Lactuca* species. Lebeda et al., (2004, 2009) elaborated a classification of *Lactuca* from taxonomic and biogeographical criteria and divided the genus into seven sections (*Lactuca* (subsection *Lactuca* and *Cyanicae* DC.), *Phaenixopus* (Cass.) Bentham, *Mulgedium* (Cass.) C.B. Clarke, *Lactucopsis* (Schultz Bip. ex Vis. Et Pančić) Rouy, *Tuberosae* Boiss., Micranthae Boiss., *Sororiae* Franchet) and two geographical groups (African and North American). Wang et al., (2013) developed a DNA-based phylogenetic tree of the *Lactuca* alliance with a hub on the Chinese center of diversity. This study fills the gap in our understanding of Asian diversity center of *Lactuca* species and related genera, exclusively for the Chinese species. However, a study of the African diversity centre of *Lactuca* species is still lacking. However, Wei et al., 2017 using chloroplast DNA reported that the endemic species of Africa probably should be treated as a new genus. The world distribution of various species of *Lactuca*is given in (Table 1).

Most of the species of *Lactuca* are xerophytes, very much adjusted to dry climatic conditions, except for some scandent, liana-like endemic species of the central African mountains (Feráková 1977). Ecologically *Lactuca* species are very diverse and occur in different habitats (Doležalová et al., 2002). Some more common European species (e.g., *L. serriola*, *L. saligna*, *L. virosa*) are usallysynanthropic (ruderal), preferring disturbed habitats (Lebeda et al., 2001). Though, some taxa (e.g., *L. aurea*, *L. quercina*, *L. biennis*, *L. sibirica*) are common in woodland habitats (Nessler 1976, Feráková 1977, Sammour 2014).

**Table 1. Different species of *Lactuca* from the World**

| S. No | Name of the species | Distribution |
|---|---|---|
| 1 | *L. acanthifolia* (Willd.) Boiss. | East Aegean Is., Greece, Kriti, Turkey. |
| 2 | *L. aculeate*Boiss. | Iran, Iraq, Lebanon-Syria, Palestine, Turkey. |
| 3 | *L. alaica*Kovalevsk. | Kirgizstan. |
| 4 | *L. alpestris* (Gand.) Rech.f. | Kriti. |
| 5 | *L. ambacensis* (Hiern) C. Jeffrey | Angola. |
| 6 | *L. attenuata* Stebbins | Burundi, Malawi, Rwanda, Uganda, Zaïre. |
| 7 | *L. azerbaijanica* Rech.f. | Iran. |
| 8 | *L. biennis* (Moench) Fernald | Alaska, Alberta, British Columbia, California, Colorado, Connecticut, Delaware, District of Columbia, Idaho, Illinois, Indiana, Iowa, Kentucky, Labrador, Maine, Manitoba, Maryland, Massachusetts, Michigan, Minnesota, Montana, New Brunswick, New Hampshire, New Jersey, New Mexico, New York, Newfoundland, North Carolina, North Dakota, Nova Scotia, Ohio, Ontario, Oregon, Pennsylvania, Prince Edward I., Québec, Rhode I., Saskatchewan, South Dakota, Tennessee, Utah, Vermont, Virginia, Washington, West Virginia, Wisconsin, Wyoming, Yukon. |
| 9 | *L. birjandica* Mozaff. | Iran. |
| 10 | *L.brachyrrhyncha* Greenm. | Mexico Central, Mexico Gulf, Mexico Northwest, Mexico Southwest. |
| 11 | *L. calophylla* C. Jeffrey | Malawi, Tanzania, Zambia. |
| 12 | *L. canadensis* L. | Alabama, Arizona, Arkansas, British Columbia, California, Colorado, Connecticut, Delaware, District of Columbia, Dominican Republic, Florida, Georgia, Haiti, Idaho, Illinois, Indiana, Iowa, Kansas, Kentucky, Louisiana, Maine, Manitoba, Maryland, Massachusetts, Michigan, Minnesota, Mississippi, Missouri, Montana, Nebraska, Nevada, New Brunswick, New Hampshire, New Jersey, New Mexico, New York, North Carolina, North Dakota, Nova Scotia, Ohio, Oklahoma, Ontario, Oregon, Pennsylvania, Prince Edward I., Québec, Rhode I., South Carolina, South Dakota, Tennessee, Texas, Utah, Vermont, Virginia, Washington, West Virginia, Wisconsin, Wyoming, Yukon. |

**Table 1. (Continued)**

| S. No | Name of the species | Distribution |
|---|---|---|
| 13 | *L. cichorioides* (Hiern) C. Jeffrey | Angola. |
| 14 | *L. corymbosa*Lawalrée | Zaïre. |
| 15 | *L. cubanguensis* (S. Moore) C. Jeffrey | Angola. |
| 16 | *L. cyprica* (Rech.f.) N. Kilian & Greuter | Cyprus. |
| 17 | *L. czerepanovii* (Kirp.) N. Kilian &Greuter | Transcaucasus. |
| 18 | *L. denaensis* N. Kilian &Djavadi | Iran. |
| 19 | *L. dissecta* D. Don | Afghanistan, East Himalaya, India, Kazakhstan, Kirgizstan, Nepal, Oman, Pakistan, Saudi Arabia, Tadzhikistan, Turkmenistan, Uzbekistan, West Himalaya, Xinjiang. |
| 20 | *L.dolichophylla*Kitam. | Afghanistan, China South-Central, East Himalaya, Myanmar, Nepal, Pakistan, Tibet, West Himalaya. |
| 21 | *L. dregeana* | Cape Provinces, Free State. |
| 22 | *L. dumicola* S. Moore | Angola. |
| 23 | *L. erostrata*Bano&Qaiser | Pakistan. |
| 24 | *L. fenzlii* N. Kilian &Greuter | Turkey. |
| 25 | *L. floridana* (L.) Gaertn. | Alabama, Arkansas, Delaware, District of Columbia, Florida, Georgia, Illinois, Indiana, Iowa, Kansas, Kentucky, Louisiana, Manitoba, Maryland, Massachusetts, Michigan, Mississippi, Missouri, Nebraska, New Jersey, New York, North Carolina, Ohio, Oklahoma, Ontario, Pennsylvania, South Carolina, South Dakota, Tennessee, Texas, Virginia, West Virginia, Wisconsin. |
| 26 | *L. formosana* Maxim. | China North-Central, China South-Central, China Southeast, Inner Mongolia, Taiwan. |
| 27 | *L. georgica*Grossh. | North Caucasus, Transcaucasus, Turkey, Turkmenistan. |

| S. No | Name of the species | Distribution |
|---|---|---|
| 28 | *L. gilanica* Mozaff. | Iran. |
| 29 | *L. glandulifera* Hook.f. | Burundi, Cameroon, Ethiopia, Gulf of Guinea Is., Ivory Coast, Kenya, Liberia, Malawi, Mozambique, Nigeria, Rwanda, Sierra Leone, Sudan, Tanzania, Uganda, Zaïre. |
| 30 | *L. glareosa* Schott & Kotschy | Turkey. |
| 31 | *L. glauciifolia* Boiss. | Afghanistan, Iran, Kirgizstan, Pakistan, Tadzhikistan, Turkey, Turkmenistan, Uzbekistan. |
| 32 | *L. gracilipetiolata* Merr. | Myanmar. |
| 33 | *L. graminifolia* Michx. | Alabama, Arizona, Bahamas, Colorado, Florida, Georgia, Guatemala, Haiti, Louisiana, Mexico Central, Mexico Gulf, Mexico Northeast, Mexico Northwest, Mexico Southeast, Mexico Southwest, Mississippi, New Jersey, New Mexico, North Carolina, South Carolina, Texas. |
| 34 | *L. haimanniana* Asch. | Libya. |
| 35 | *L. hazaranensis* Djavadi & N. Kilian | Iran. |
| 36 | *L. hirsute*Muhl. ex Nutt. | Alabama, Arkansas, Connecticut, District of Columbia, Georgia, Illinois, Indiana, Kentucky, Louisiana, Maine, Maryland, Massachusetts, Michigan, Mississippi, Missouri, New Hampshire, New Jersey, New York, North Carolina, Nova Scotia, Ohio, Ontario, Pennsylvania, Prince Edward I., Québec, Rhode I., South Carolina, Tennessee, Texas, Vermont, Virginia, West Virginia. |
| 37 | *L. hispida* DC. | Albania, Bulgaria, East Aegean Is., Greece, Krym, North Caucasus, Palestine, Romania, Transcaucasus, Turkey, Ukraine, Yugoslavia. |
| 38 | *L. hispidula* B. Fedtsch. | Turkmenistan. |
| 39 | *L. homblei* De Wild. | Zambia, Zaïre. |
| 40 | *L. imbricate* Hiern | Angola, Malawi, Mozambique, Tanzania, Zambia, Zaïre, Zimbabwe. |
| 41 | *L. indica* L. | Amur, Assam, China North-Central, China South-Central, China Southeast, Chita, East Himalaya, Hainan, Japan, Jawa, Khabarovsk, Korea, Laos, Manchuria, Nansei-shoto, Philippines, Primorye, Taiwan, Thailand, Tibet, Vietnam. |

**Table 1. (Continued)**

| S. No | Name of the species | Distribution |
|---|---|---|
| 42 | *L. inermis* Forssk. | Angola, Benin, Burkina, Cameroon, Cape Provinces, Central African Repu, Chad, Congo, Eritrea, Ethiopia, Free State, Gulf of Guinea Is., Ivory Coast, Kenya, KwaZulu-Natal, Lesotho, Madagascar, Mali, Mozambique, Namibia, Northern Provinces, Oman, Saudi Arabia, Sudan, Swaziland, Tanzania, Togo, Uganda, Yemen, Zambia, Zaïre, Zimbabwe. |
| 43 | *L. intricate* Boiss. | Albania, East Aegean Is., Greece, Turkey. |
| 44 | *L. kanitziana* Martelli | Borneo. |
| 45 | *L. kemaliya* Yild. | Turkey. |
| 46 | *L. kirpicznikovii* (Grossh.) N. Kilian & Greuter | Transcaucasus. |
| 47 | *L. klossii* S. Moore | Vietnam. |
| 48 | *L. kochiana* Beauverd | Transcaucasus, Turkey. |
| 49 | *L. laevigata* DC. | Jawa, New Guinea. |
| 50 | *L. lasiorhiza* (O. Hoffm.) C. Jeffrey | Cameroon, Central African Repu, Ghana, Malawi, Nigeria, Tanzania, Togo, Zambia, Zaïre, Zimbabwe. |
| 51 | *L. leucoclada* Rech.f. &Tuisl | Afghanistan. |
| 52 | *L. longespicata* De Wild. | Tanzania, Zambia, Zaïre. |
| 53 | *L. longidentata* Moris ex DC. | Sardegna. |
| 54 | *L. ludoviciana* (Nutt.) Riddell | Arizona, Arkansas, British Columbia, California, Colorado, Idaho, Illinois, Indiana, Iowa, Kansas, Kentucky, Louisiana, Manitoba, Mexico Northeast, Minnesota, Montana, Nebraska, New Mexico, North Dakota, Oklahoma, Oregon, Saskatchewan, South Dakota, Texas, Utah, Washington, Wisconsin, Wyoming. |
| 55 | *L. malaissei* Lawalrée | Zaïre. |
| 56 | *L. mansuensis* Hayata | Taiwan. |
| 57 | *L. marunguensis* Lawalrée | Zaïre. |

| S. No | Name of the species | Distribution |
|---|---|---|
| 58 | *L. microcephala* DC. | Iran, Iraq, Lebanon-Syria, Transcaucasus, Turkey. |
| 59 | *L. microsperma* K. Schum. | New Guinea. |
| 60 | *L. morssii* B. L. Rob. | Illinois, Maine, Massachusetts, Michigan, New York, Ohio, Virginia. |
| 61 | *L. muralis* (L.) E. Mey. | Albania, Algeria, Austria, Baltic States, Belarus, Belgium, Bulgaria, Central European Rus, Corse, Czechoslovakia, Denmark, East European Russia, Finland, France, Germany, Great Britain, Greece, Hungary, Ireland, Italy, Krym, Morocco, Netherlands, North Caucasus, Northwest European R, Norway, Poland, Romania, Sardegna, Sicilia, South European Russi, Spain, Sweden, Switzerland, Transcaucasus, Tunisia, Turkey, Turkey-in-Europe, Ukraine, Yugoslavia. |
| 62 | *L. mwinilungensis* G. V. Pope | Zambia, Zaïre. |
| 63 | *L. orientalis*Boiss. | Afghanistan, Egypt, Iran, Iraq, Kazakhstan, Kirgizstan, Lebanon-Syria, Oman, Pakistan, Palestine, Saudi Arabia, Sinai, Tadzhikistan, Transcaucasus, Turkey, Turkmenistan, Uzbekistan, West Himalaya, Xinjiang. |
| 64 | *L. oyukludaghensis* (Parolly) N. Kilian &Parolly | Turkey. |
| 65 | *L. pakistanica* T. Akhtar &Chaudhri | Pakistan. |
| 66 | *L. palmensis* Bolle | Canary Is. |
| 67 | *L. paradoxa* Sch. Bip. ex A. Rich. | Ethiopia, Kenya, Malawi, Tanzania, Uganda, Zaïre. |
| 68 | *L. parishii* Craib ex Hosseus | China South-Central, Myanmar, Thailand, Vietnam. |
| 69 | *L. perennis* L. | Albania, Austria, Baltic States, Belgium, Bulgaria, Czechoslovakia, France, Germany, Greece, Hungary, Italy, Poland, Romania, Spain, Switzerland, Ukraine, Yugoslavia. |
| 70 | *L. petrensis* Hiern | Angola. |
| 71 | *L. piestocarpa* (Boiss.) Sennikov | Iran. |
| 72 | *L. polyclada* Boiss. | Iran. |
| 73 | *L. praecox* R. E. Fr. | Malawi, Tanzania, Zambia, Zaïre. |

**Table 1. (Continued)**

| S. No | Name of the species | Distribution |
|---|---|---|
| 74 | *L. praevia* C. D. Adams | Burkina, Guinea. |
| 75 | *L. pulchella* DC. | Alaska, Alberta, Arizona, British Columbia, California, Colorado, Idaho, Illinois, Indiana, Iowa, Kansas, Louisiana, Maine, Manitoba, Mexico Northeast, Mexico Northwest, Michigan, Minnesota, Missouri, Montana, Nebraska, Nevada, New Mexico, New York, North Dakota, Northwest Territorie, Ohio, Oklahoma, Ontario, Oregon, Pennsylvania, Québec, Saskatchewan, South Dakota, Texas, Utah, Washington, Wisconsin, Wyoming. |
| 76 | *L. pumila* Rech.f. & Tuisl | Afghanistan. |
| 77 | *L. quercina* L. | Albania, Austria, Bulgaria, Central European Rus, Czechoslovakia, East European Russia, France, Germany, Hungary, Iran, Italy, Krym, North Caucasus, Romania, South European Russi, Sweden, Transcaucasus, Turkey, Ukraine, West Himalaya, Yugoslavia. |
| 78 | *L. raddeana* Maxim. | Amur, China North-Central, China South-Central, China Southeast, Japan, Khabarovsk, Korea, Kuril Is., Manchuria, Primorye, Sakhalin, Vietnam. |
| 79 | *L. rechingeriana* (Tuisl) N. Kilian &Greuter | Iran, Iraq, Turkey. |
| 80 | *L. reviersii* Litard. &Maire | Morocco. |
| 81 | *L. rostrate* Kuntze | Himalaya. |
| 82 | *L. rosularis* Boiss. | Iran, Turkmenistan. |
| 83 | *L. saligna* L. | Albania, Algeria, Austria, Azores, Baleares, Belgium, Bulgaria, Central European Rus, Corse, Cyprus, Czechoslovakia, East Aegean Is., East European Russia, Egypt, France, Germany, Great Britain, Greece, Hungary, Iran, Iraq, Italy, Kriti, Krym, Lebanon-Syria, Morocco, Netherlands, North Caucasus, Palestine, Poland, Portugal, Romania, Sardegna, Saudi Arabia, Sicilia, Sinai, South European Russi, Spain, Switzerland, Transcaucasus, Tunisia, Turkey, Turkey-in-Europe, Ukraine, Yugoslavia. |
| 84 | *L. sativa* L. | Iraq. |

| S. No | Name of the species | Distribution |
|---|---|---|
| 85 | *L. scarioloides* Boiss. | Afghanistan, Iran, Iraq, Turkey. |
| 86 | *L. schulzeana* Büttner | Angola, Cameroon, Congo, Zambia, Zaïre. |
| 87 | *L. schweinfurthii* Oliv. &Hiern | Angola, Burundi, Cameroon, Malawi, Sudan, Tanzania, Zambia, Zaïre. |
| 88 | *L. serriola* L. | Afghanistan, Albania, Algeria, Altay, Austria, Azores, Baleares, Baltic States, Belarus, Belgium, Bulgaria, Canary Is., Central European Rus, Corse, Cyprus, Czechoslovakia, Denmark, East Aegean Is., East European Russia, Egypt, Eritrea, Ethiopia, Finland, France, Germany, Great Britain, Greece, Hungary, Iran, Iraq, Ireland, Italy, Kazakhstan, Kirgizstan, Kriti, Krym, Lebanon-Syria, Libya, Madeira, Mongolia, Morocco, Netherlands, North Caucasus, North European Russi, Northwest European R, Norway, Pakistan, Palestine, Poland, Portugal, Romania, Sardegna, Saudi Arabia, Sicilia, Sinai, Somalia, South European Russi, Spain, Sudan, Sweden, Switzerland, Tadzhikistan, Transcaucasus, Tunisia, Turkey, Turkey-in-Europe, Turkmenistan, Tuva, Ukraine, Uzbekistan, West Himalaya, West Siberia, Xinjiang, Yemen, Yugoslavia. |
| 89 | *L. setosa* Stebbins ex C. Jeffrey | Malawi, Tanzania, Zambia, Zaïre, Zimbabwe. |
| 90 | *L. sibirica* (L.) Benth. ex Maxim. | Altay, Amur, Baltic States, Belarus, Buryatiya, Central European Rus, China North-Central, Chita, East European Russia, Finland, Inner Mongolia, Irkutsk, Japan, Kamchatka, Kazakhstan, Khabarovsk, Korea, Krasnoyarsk, Kuril Is., Magadan, Manchuria, Mongolia, North European Russi, Northwest European R, Norway, Primorye, Qinghai, Sakhalin, Sweden, Tuva, Ukraine, West Siberia, Xinjiang, Yakutskiya. |
| 91 | *L. singularis* Wilmott | Nevada, Spain. |
| 92 | *L. songeensis* C. Jeffrey | Tanzania, Zaïre. |
| 93 | *L. spinidens* Nevski | Turkmenistan, Uzbekistan |
| 94 | *L. stebbinsii* N. Kilian | Angola. |
| 95 | *L. stipulata* Stebbins | Rwanda. |
| 96 | *L. Lactuca* takhtadzhianii Sosn. | Iran, Transcaucasus. |

**Table 1. (Continued)**

| S. No | Name of the species | Distribution |
|---|---|---|
| 97 | *L. tatarica* C. A. Mey. | Afghanistan, Altay, Baltic States, Bulgaria, Buryatiya, Central European Rus, China North-Central, China Southeast, Czechoslovakia, East European Russia, Inner Mongolia, Iran, Irkutsk, Kazakhstan, Kirgizstan, Krasnoyarsk, Krym, Manchuria, Mongolia, North Caucasus, North European Russi, Northwest European R, Pakistan, Qinghai, Romania, South European Russi, Tadzhikistan, Tibet, Transcaucasus, Turkey, Turkey-in-Europe, Turkmenistan, Tuva, Ukraine, Uzbekistan, West Himalaya, West Siberia, Xinjiang. |
| 98 | *L. tenerrima*Pourr. | Baleares, France, Morocco, Spain. |
| 99 | *L. tetrantha* B. L. Burtt & P. H. Davis | Cyprus. |
| 100 | *L. tinctociliata* I. M. Johnst. | Angola. |
| 101 | *L. triangulata* Maxim. | Amur, China North-Central, Japan, Khabarovsk, Korea, Manchuria, Primorye, Vietnam. |
| 102 | *L. tuberosa* Jacq. | Bulgaria, Cyprus, East Aegean Is., Greece, Iran, Iraq, Kriti, Krym, Lebanon-Syria, North. Caucasus, Oman, Palestine, Transcaucasus, Turkey, Ukraine, Yugoslavia. |
| 103 | *L. tysonii* (E. Phillips) C. Jeffrey | Cape Provinces, KwaZulu-Natal. |
| 104 | *L. ugandensis* C. Jeffrey | Burundi, Cameroon, Uganda, Zaïre. |
| 105 | *L. undulate* Ledeb. | Afghanistan, East Himalaya, Iran, Iraq, Kazakhstan, Kirgizstan, Lebanon-Syria, Mongolia, Nepal, Pakistan, Palestine, Sinai, Tadzhikistan, Transcaucasus, Turkey, Turkmenistan, Uzbekistan, West Himalaya, Xinjiang. |
| 106 | *L. viminea* (L.) J. Presl & C. Presl | Albania, Austria, Baleares, Bulgaria, Canary Is., Corse, Cyprus, Czechoslovakia, East Aegean Is., East European Russia, France, Germany, Great Britain, Greece, Hungary, Iran, Iraq, Italy, Kriti, Krym, Lebanon-Syria, North Caucasus, Pakistan, Palestine, Portugal, Romania, Sardegna, Sicilia, South European Russi, Spain, Switzerland, Transcaucasus, Turkey, Turkey-in-Europe, Turkmenistan, Ukraine, West Himalaya, Yugoslavia. |

| S. No | Name of the species | Distribution |
|---|---|---|
| 107 | *L. virosa* L. | Algeria, Austria, Baleares, Belgium, Corse, Czechoslovakia, France, Germany, Great Britain, Greece, Hungary, Ireland, Italy, Madeira, Morocco, Netherlands, Poland, Portugal, Romania, Sardegna, Sicilia, Spain, Switzerland, Yugoslavia. |
| 108 | *L. watsoniana* Trel. | Azores. |
| 19 | *L. winkleri* Kirp. | Central Asia. |
| 110 | *L. yemensis* Deflers | Yemen. |
| 111 | *L. zambeziaca* C. Jeffrey | Angola, Zambia, Zaïre. |

Most of *Lactuca* species are calciphilous plants and settle limestone and dolomite areas, generally rocky slopes (*L. perennis*, *L. viminea*, *L. graeca*, *L. tenerrima*) (Lopez & Jimenez 1974). *L. tatarica* and *L. acanthifolia* grow on cliffs at the seashore, but *L. tatarica* expands to Asia and Central and North Europe as a weedy species covering unfertile salt substrata as well (Sammour 2014, Jehlík 1998).

## Origin

Zohary (1991) assumed that *L. dregeana* DC. originated and spread in the region from Eurasia to South Africa. Moreover, he pointed out that much less attention has been paid to the Southwest Asian species about the origin of cultivated lettuce. *L. aculeata* Bois. & Kotschy ex Bois., *L. scarioloides* Boiss. and *L. Azerbaijanica* Rech. appeared in the centre of origin (Southwest Asia) of lettuce. *L. serriola* is generally believed native to Europe, western Asia and northern Africa (Sammour 2014). It is likely originated on the Mediterranean rim on rocky wasteland or woodland clearings. In general, European *Lactuca* species have a center of origin in the Mediterranean region (Lebeda et al., 2001).

There are several different opinions about the center of origin of cultivated lettuce (*L. sativa*). As indicated by Lindqvist (1960), it most likely originated from Egypt. As per Vavilov, cultivated lettuce originated in the Mediterranean area (Ryder 1986). This is also the view of Harlan (1992). Boukema et al., (1990) stated that the domestication of lettuce took place in Southwest Asia in the region between Egypt and Iran. Rulkens (1987) stated that cultivated lettuce has its origin in the Kurdistan-Mesopotamia area and not in Egypt. Kesseli et al., (1991) announced that Lindqvist (1960) and De Candolle (1959) put several views on the origin of cultivated lettuce.

These were that: (1) lettuce arose from wild forms of *L. sativa*; (2) it is originated directly from wild *L. serriola*; and (3) lettuce is the result of hybridizations between different species. The by and large held view was hypothesis 2; there are no extant, truly wild *L. sativa* populations and *L. serriola* was thought to encompass fully the variation of cultivated lettuce (Ryder & Whitaker 1986). Nonetheless, Lindqvist (1960) concluded that hypothesis 3 was the most conceivable since he found two morphological characters of *L. sativa* (pointed leaf apex and spotted anthocyanin on leaves) not known in *L. serriola* but present in *L. saligna* and *L. virosa*, respectively. Ryder and Whitaker (1976) further expounded on the hybridization hypothesis of Lindqvist, pointing out that relationships between progenitor and derivative species may be difficult to define.

Lettuce comes in a variety of colors, sizes, and shapes and as a result of this diversity, lettuces can be grouped by their types. A type is a group of morphologically similar cultivars. A type can be additionally subdivided into subtypes which share more morphological and genetic similarities. A cultivar is a variety selected for desirable traits for cultivation. A variety is a taxonomic rank below species and subspecies. Even though the fact that there have been different classification systems proposed by different groups of researchers over the years (Rodenburg 1960, Lebedaet al., 2007), there is no standardized classification system, due to high genetic and morphological diversity among lettuce cultivars. According to Mou (2008), there are six main lettuce types based on leaf shape, size, texture, head formation, and stem type. They are (1) crisphead lettuce (var. *capitata* L. *nidus jaggeri* Helm), (2) butterhead lettuce (var. *capitata* L. *nidus tenerrima* Helm), (3) romaine or cos lettuce (var. *longifolia* Lam., var. *romana* Hort. In Bailey), (4) leaf or cutting lettuce (var. *acephala* Alef., syn. var. *secalina* Alef., syn. var. *crispa* L.), (5) stem or stalk (Asparagus) lettuce (var. *angustana* Irish ex Bremer, syn. var. *asparagina* Bailey, syn. L. *angustana* Hort. In Vilm.), and (6) Latin lettuce. Scientific names of varieties for each lettuce type are also included according to Lebeda et al., (2007) and Křístková et al., (2008).

In 2014, the production of head types, romaine, and leaf type lettuces in the US were 54.1%, 30.1% and 15.8%, respectively (USDA 2014).

## Economic, Edible and Medicinal Importance

Several species of the genus *Lactuca* L. have been cultivated since ancient times for their economic and medicinal importance. Cultivated lettuce (*L. sativa*) is the most important leafy salad vegetable in the world (Sammour 2014) and is a good source of high dietary fiber (1.1 g/100 g, FW), vitamin A (166 μg/100 g, FW), folate or vitamin B9 (73 μg/100 g, FW), vitamin C (4 mg/100 g, FW), vitamin K (24 μg/100 g), and phenolic compounds. Lettuce are very low in calories (10 kcal [60 kJ]/100 g, FW), it is often prescribed for weight loss programs (Niederwieser 2001). Phenolic compounds in plants are responsible for antioxidant scavenging properties. Carotenoids possess the antioxidant capacity and vitamin C, the water-soluble antioxidant, also shows antioxidant properties (Lópezet al., 2014). Colorful lettuce types have increased in recent years and darker lettuce types (romaine, green leaf and red leaf) were reported to be better sources for vitamin A, niacin, riboflavin, thiamine, Ca, Fe, K, Mn and Se (Bunning et al., 2010). The higher β-carotene and lutein content in lettuce is associated with reducing the risk of cancers, cataracts and heart disease and stroke (USDA 2004). The oil of the oilseed group, with high vitamin E content, is utilized for human consumption. *L. serriola* and other wild *Lactuca* species can be eaten as a salad, despite they have something of a bitter taste. Some species of the genus *Lactuca* L. are rich in a milky sap that flows freely from any wounds in the plant. This sap contains *Lactuca*rium which is utilized in medicine for its anodyne, antispasmodic, digestive, diuretic, narcotic, aphrodisiac, soporific and sedative properties. It is taken internally in the treatment of insomnia, anxiety, neuroses, hyperactivity in children, dry coughs, whooping cough, rheumatic pain (Bown 1995).

The sap has additionally been applied externally in the treatments of warts. In the Unani system of medicine, Kahu is equated with *L. sativa* Linn (Anonymous 1997). The color, size, texture, and taste are important parameters for successful marketing of lettuce and these factors determine the market price and consumer preference. Lettuce is marketed as a whole product or as fresh cuts.

In the ancient Egypt period, Lettuce was first cultivated for the production of oil from its seed (Ali et al., 2016). Kahu is an herb and it is of two varieties one is wild Sahrai and the other is garden Bustani (Kabiruddin&Makhzan-ul-Mufradat 2007). SahraiKahu is of two types, first one has more wide leaves, the height of the plant is 7.28 m, sweet, soft, the stem is thin, flowers are white and in India, it is cultivated in the winter session and Arab cultivated in the summer season. The second one, firangistani is again two types first one has green, sweet and soft leaves and the second one have blue margin leaves (Ali et al., 2016). The leaves of garden Kahu are used as vegetables and seeds are used as medicine. Exudate is obtained from the Kahu plant, known as AfyunKahu. It is also used for the medicinal purposes (Kabiruddin&Makhzan-ul-Mufradat 2007).

Lettuce is preferably grown in the hydroponic system, and the growth rate of a hydroponic plant is faster and the yield is higher than a soil-grown plant produced under the same conditions. Also, the hydroponic system uses less water due to the constant reuse of nutrient solution and also reduces the risk of soil-borne diseases. Phytochemical contents in fresh produce can be improved when osmotic or salt stress is created with a high nutrient solution electrical conductivity (EC) in the root zone in a hydroponic system (Cornish 1992, De Pascale et al., 2001).

Some viral, bacterial and fungal pathogens, such as yellows virus, Turnip mosaic potyvirus, *Microdochium*, *Rhizomonas* (corky root disease), *Bremia* (downy mildew), and *Erysiphe* (Galea & Price 1988) are one of the most important problems affecting lettuce. Likewise, some insect pathogens such as *Bremisia* spp. and *Trichoplusiani* is the only insects reported as causing damage to leaves of *L. serriola* and *L. sativa* (Coudriet et al., 1986, Dussourd 1997). The utilization of pesticides to control these devastating diseases is harmful to human and the environment. Thus, attention has been paid to genetic resources to find the genes conferring resistance, and the genes conferring high and good agronomic traits, including oil content in oil-producing species. The biggest producer of lettuce in the world is China, which produces mainly stem lettuce (*L. sativa* L. var. *angustana*) not commonly consumed in the US or Western Europe (Křístková 2008). The US and Western Europe contribute about 22% and 13% of the total lettuce production worldwide, respectively (Mou 2008). In the US, lettuce ranks as the 3rd most consumed vegetable (USDA 2015). A cohort study conducted in the US consisting of male and female members (n = 186, 916) representing five ethnic groups reported that lettuce was the most consumed vegetable (6.0–9.9%) in all ethnic and gender groups except for African-American women and Mexican-born Latino men and women (Sharma et al., 2014). Even though the lettuce is a popularly consumed vegetable, it has not been viewed as nutritional food, because due to its high water content (95%). However, the nutrient composition can be equivalent to other vegetables known as "nutritious" depending on lettuce type.

## Biological Activities

Wild *Lactuca* species have been little studied chemically. Until now, some 18 wild *Lactuca* species originated from Europe, Asia, and North America have been investigated yielding mainly sesquiterpene lactones (Michalska et al., 2009) and phenolic compounds (Stojakowska et al., 2013). Earlier research shows that the *Lactuca* species display a wide range of biological activities, such as anticancer, anti-inflammatory and others like antioxidant, antidiabetic, antiviral and antimicrobial activity. An overview of the modern pharmacological assessments carried out on these species has been described in the following sections.

## Table 2. An Overview of the Modern Pharmacological Assessments

| Biological activities | *Lactuca* species | Reference |
|---|---|---|
| Anti-inflammatory | *L. sativa* | Sayyahet al., (2004), Pepe et al., (2015), Adessoet al., (2016), Złotek and Świeca 2016. |
| | *L. sativa* var. crispa | Zekkoriet al., (2018). |
| Antioxidant activity | *L. sativa* | Caldwell 2003, Altunkaya et al., (2009), Oh et al., (2011), Al Nomaani et al., (2013), Durazzo et al., (2014), Pepe et al., (2015), Adesso et al., (2016), Mampholo et al., (2016), Złotek and Świeca 2016, Qin et al., 2018, Algfri et al., (2019). |
| | *L. sativa* var. crispa | Zekkori et al., (2018). |
| | *L. tuberose* | Stojakowska et al., (2013). |
| | *L. indica* | Wang et al., (2003), Choi et al., 2016. |
| | *L. runcinata* | Devi and Muthu 2014. |
| | *L. serriola* | Urmila et al., 2013. |
| | *L. orientalis* | Jaradat et al., 2017. |
| | *L. sativa* var longifolia | Edziri et al., 2011. |
| Antianxiety property | *L. sativa* | Harsha and Anilakumar 2012. |
| Superoxide dismutase activity | *L. sativa* | Ruiz-Lozano et al., (1996). |
| Phytotoxic property | *L. sativa* | Gui et al., (2015). |
| Anxiolytic property | *L. sativa* | Harsha and Anilakumar 2012. |
| Uroepithelial infection | *L. indica* | Lüthje et al., 2011. |
| Hepatoprotective activity | *L. indica* | Kim et al., 2007. |
| Anti-diabetic activity | *L. indica* | Choi et al., 2016. |
| Anti-diabetic activity | *L. scariola* | Chadchan et al., 2016. |
| Anticancer activity | *L. serriola* | Elsharkawy and Alshathly 2013. |
| Anti-Tumor Potential | *L. sativa* | Qin et al., 2018. |

**Table 2. (Continued)**

| Biological activities | *Lactuca* species | Reference |
|---|---|---|
| Anti-leukaemic effects | *L. sativa* | Gridling et al., 2010. |
| Antiviral activity | *L. sativa* var Longifolia | Edziri et al., 2011. |
| | *L. sativa* | Matvieieva et al., 2012. |

**Table 3. Antimicrobial Study**

| Species | Microorganism | Reference |
|---|---|---|
| *Lactucaserriola* | *Staphylococcus aureus*, *S. epidermedis*, *S. saprophyticus*, *Klebsilla*, *Serratia*, *Proteus*, *Escherichiacoli*, *Pseudomonas*, *Provedenatia* | Al-Marzoqiet al., 2015. |
| *L. scariola* | Bacterial species<br>(+) *Bacilluscoagulas*<br>(-) *Escherichiacoli*<br>(+) *S.aureus*<br>Fungal species<br>*Penicilliumdigitatum*<br>*Aspergillusniger*<br>*Trichodermaviride* | Yadava and Jharbade 2007. |
| *L. orientalis* | Bacterial species<br>*S. aureus* (ATCC 25923)<br>*E. coli* (ATCC25922)<br>*Pseudomonasaeruginosa* (ATCC27853)<br>*S. aureus* (MRSA Positive)<br>*Enterococcusfaecium* (ATCC 700221)<br>*Shigellasonnei* (ATCC 25931)<br>*Enterobactercloacae* (clinical)<br>*Klebsiellapneumoniae* (clinical)<br>Fungal species<br>*Candidaalbicans* (ATCC90028)<br>*Epidermophytonfloccosum* (ATCC52066). | Jaradatet al., 2017. |
| *L. sativa* var longifolia | Gram-negative bacteria<br>*E. coli* ATCC25922<br>*E. coli* (clinical isolated)<br>*Klebsiella pneumoniae* (clinical isolated)<br>*Enterobacter cloacae* (clinical isolated)<br>*Serratiamarcescens* (clinical isolated) | Edziriet al., 2011. |

| Species | Microorganism | Reference |
|---|---|---|
| | *Acinetobacter baumannii* (clinical isolated) | |
| | Gram-positive bacteria | |
| | *Bacillus subtilus* ATCC 6633 | |
| | *S. aureus* (clinical isolated) | |
| | *E. faecalis* ATCC 29212 | |
| | *E. faecium* (clinical isolated) | |
| | *Corynebacterium* spp. (clinical isolated) | |
| *L. sativa* | Bacterial species | Zdravkovic et al., 2012. |
| | *S. aureus* ATCC 25923 | |
| | *K. pneumoniae* ATCC13883 | |
| | *E. coli* ATCC 25922 | |
| | *Proteus vulgaris* ATCC13315 | |
| | *P. mirabilis* ATCC14153 | |
| | *B. subtilis* ATCC6633 | |
| | Fungal species | |
| | *C. albicans* ATCC10231 | |
| | *A. niger* ATCC16404 | |

## CONCLUSION

The data summarized in the article show that the genus *Lactuca* includes a relatively large number of species distributed in most of the world's important biogeographical and phytogeographical regions. Several species are often grown as a leaf vegetable and were first cultivated by the Egyptians. Findings from *in vitro* studies indicated that members of the genus have a potential effect against different cancer lines, tumor, viral and microbial strains. They also showed significant anti-inflammatory, analgesic, and hepatoprotective activities. The most abundant bioactive secondary metabolites in members of the genus are *Lactucarium* and terpenoids which are also mentioned as responsible for the cytotoxic effect observed. We believe this review will provide summarized information to the scientific community working on the genus.

## REFERENCES

[1] Adesso S, Pepe G, Sommella E, Manfra M, Scopa A, Sofo A, Tenore G.C, Russo M, Di Gaudio F, Autore G, and Campiglia P. 2016. "Anti-inflammatory and antioxidant activity of polyphenolic extracts from *Lactuca sativa* (var. *Maravilla de Verano*) under different farming methods." *Journal of the Science of Food and Agriculture* 96: 4194–4206.

[2] Al Nomaani R.S.S, Hossain M.A, Weli A.M, Al-Riyami Q, and Al-Sabahi J.N. 2013. "Chemical composition of essential oils and in vitro antioxidant activity of fresh and dry leaves crude extracts of medicinal plant of *Lactuca sativa* L. native to Sultanate of Oman." *Asian Pacific journal of tropical biomedicine* 3: 353–357.

[3] Algfri S.K, Kaid A.A, and Munaiem R.T. 2019. "Phytochemical and antioxidant studies of some green leafy vegetables consumed in Yemen as salad". *Journal of Medicinal Plants* 7: 22–28.

[4] Ali W, Ahmad H.A, Mohammad A, and Nasir A. 2016. "Tukhmekahu (*Lactuca sativa* Linn): Pharmacological and phytochemical profile and uses in unani medicine." *Journal of Pharmaceutical and Scientific Innovation* 5: 1–4.

[5] Al-Marzoqi A.H, Hussein H.J, and Al-Khafaji N.M.S. 2015. "Antibacterial activity of the crude phenolic, alkaloid and terpenoid compounds extracts of *Lactucaserriola* L. on human pathogenic bacteria." *Chemistry and Materials Research* 7: 8–10.

[6] Altunkaya A, Becker E.M, Gökmen V, and Skibsted L.H. 2009. "Antioxidant activity of lettuce extract (*Lactuca sativa*) and synergism with added phenolic antioxidants." *Food chemistry* 115: 163–168.

[7] Anonymous. 1997. *Standardization of single drugs of Unani medicine*. Part III. New Delhi: CCRUM, Ministry of H & FW, Govt. of India.

[8] Boukema, I.W., Hazekamp, T. and van Hintum, T.J., 1990. *The CGN lettuce collection*. Centre for Genetic Resources.

[9] Bown, D. 1995. *Encyclopaedia of herbs and their uses. Dorling Kindersley*. London ISBN 0-7513-020-31.

[10] Bremer, K., Anderberg, A.A., Karis, P.O., Nordenstam, B., Lundberg, J. and Ryding, O. 1994. *Asteraceae: Cladistic and classification*. TimberPress, (Portland).

[11] Bunning M.L, Kendall P.A, Stone M.B, Stonaker F.H. and Stushnoff C. 2010. "Effects of seasonal variation on sensory properties and total phenolic content of 5 lettuce cultivars." *Journal of food science* 75: 156−161.

[12] Caldwell, C.R. 2003. "Alkylperoxyl radical scavenging activity of red leaf lettuce (*Lactuca sativa* L.) phenolics." *Journal of agricultural and food chemistry* 51: 4589−4595.

[13] Chadchan K.S, Jargar J.G, and Das S.N. 2016. "Anti-diabetic effects of aqueous prickly lettuce (*Lactucas cariola* Linn.) leaves extract in alloxan-induced male diabetic rats treated with nickel (II)." *Journal of basic and clinical physiology and pharmacology* 27: 49−56.

[14] Choi C.I, Eom H.J, and Kim K.H. 2016. "Antioxidant and α-glucosidase inhibitory phenolic constituents of *Lactuca indica* L." *Russian Journal of Bioorganic Chemistry* 42: 310−315.

[15] Cornish P.S. 1992. "Use of high electrical conductivity of nutrient solution to improve the quality of salad tomatoes (*Lycopersicone sculentum*) grown in hydroponic culture." *Australian Journal of Experimental Agriculture*, 32: 513−520.

[16] Coudriet D.L, Meyerdirk D.E, Prabhaker N, and Kishaba A.N. 1986. "Bionomics of sweet potato white fly (Homoptera: Aleyrodidae) on weed hosts in the Imperial Valley, California." *Environmental Entomology* 15: 1179−1183.

[17] De Candolle, A. 1959. "*Origin of Cultivated Plants. Hafner Publishing Company*. New York and London.

[18] Devi J.A.I, and Muthu A.K. 2014. "*In vitro* anti oxidant activity and total phenolic content of ethanolic extract of whole plant of *Lactucaruncinata* (DC)." *3rd International conference on medical, biological and pharmaceutical sciences* (ICMBPS'2014).

[19] Doležalová I, Křístková E, Lebeda A, and Vinter V. 2002. "Description of morphological characters of wild *Lactuca* L. spp. genetic resources (English-Czech version)." *Horticultural Science* (Prague), 29: 56–83.

[20] Durazzo A, Azzini E, Lazzé M.C, Raguzzini A, Pizzala R, Maiani G, Palomba L, and Maiani G. 2014. "Antioxidants in Italian head lettuce (*Lactuca sativa* var. *capitata* L.) grown in organic and conventional systems under greenhouse conditions." *Journal of Food Biochemistry* 38: 56–61.

[21] Dussourd D.E. 1997. "Plant exudates trigger leaf-trenching by cabbage loopers, Trichoplusiani (Noctuidae)."*Oecologia* 112: 362–369.

[22] Edziri H.L, Smach M.A, Ammar S, Mahjoub M.A, Mighri Z, Aouni M, and Mastouri M, 2011. "Antioxidant, antibacterial, and antiviral effects of *Lactuca sativa* extracts." *Industrial Crops and Products* 34: 1182–1185.

[23] Elsharkawy E, and Alshathly M. 2013. "Anticancer activity of *Lactucaserriola* growing under dry desert condition of Northern Region in Saudi Arabia." *Journal of Nature and Science* 3: 5–18.

[24] Feráková, V. 1977. *The genus Lactuca L. in Europe*. Univerzita Komenského, Bratislava.

[25] Frietema F.T. 1994. "The systematic relationship of *Lactuca sativa* and *Lactucaserriola*, in relation to the distribution of prickly lettuce." *Actabotanicaneerlandica* 43: 79.

[26] Galea V.J, and Price T.V. 1988. "Infection of lettuce by *Microdochium panattonianum*." *Transactions of the British Mycological Society* 91: 419–425.

[27] Gridling M, Popescu R, Kopp B, Wagner K.H, Krenn L, and Krupitza G. 2010. "Anti-leukaemic effects of two extract types of *Lactuca sativa* correlate with the activation of Chk2, induction of p21, downregulation of cyclin D1 and acetylation of α-tubulin." *Oncology reports* 23: 1145–1151.

[28] Gui X, Zhang Z, Liu S, Ma Y, Zhang P, He X, Li Y, Zhang J, Li H, Rui Y, and Liu L. 2015. "Fate and phytotoxicity of $CeO_2$

nanoparticles on lettuce cultured in the potting soil environment." *PloS one* 10: p.e0134261.

[29] Harlan J. 1992. 2nd Edition, American Society of Agronomy, Inc. Crop Science Society of America, Madison, Wisconsin, USA.

[30] Harsha S.N, and Anilakumar K.R. 2012. "Effects of *Lactuca sativa* extract on exploratory behavior pattern, locomotor activity and anxiety in mice." *Asian Pacific Journal of Tropical Disease* 2: 475−479.

[31] Hartman Y, Hooftman D.A, Uwimana B, Schranz M.E, Van de Wiel C.C, Smulders M.J, Visser R.G, Michelmore R.W. and Van Tienderen P.H. 2014. "Abiotic stress QTL in lettuce crop–wild hybrids: comparing greenhouse and field experiments." *Ecology and evolution* 4: 2395−2409.

[32] Hartman Y, Hooftman D.A.P, Schranz M.E, and van Tienderen P.H. 2012. QTL analysis reveals the genetic architecture of domestication traits in Crisphead lettuce. *Genetic Resources and Crop Evolution* 60: 1487−1500.

[33] Heywood, V. H. 1978. *Flowering plants of the world*. Oxford Univ. Press, Oxford.

[34] Jaradat N, AlMasri M, and Zaid A.N. 2017. "Pharmacological and phytochemical screening of Palestinian traditional medicinal plants *Erodium laciniatum* and *Lactuca orientalis*." *Journal of Complementary and Integrative Medicine*, 15(1).

[35] Jeffrey C. 1966. *Notes on Compositae*: *I. Kew Bull* 18: 427−486.

[36] Jehlík V. (Ed.) 1998. *Alien expansive weeds of the Czech Republic and the Slovak Republic*. Academia, Praha (Czech Republic).

[37] Jeuken M.J, Pelgrom K, Stam P, and Lindhout P. 2008. "Efficient QTL detection for nonhost resistance in wild lettuce: backcross inbred lines versus F2 population." *Theoretical and Applied Genetics* 116: 845−8573.

[38] Judd W.S, C. S. Campbell, E. A. Kellogg & P. F. Stevens. 1999. *Plant systematics: A phylogenetic approach.* Sinauer Assoc., Sunderland, MA.

[39] Judd, W.S., Campbell, C., Kellogg, E.A., Stevens, P.F., and Donoghue M.J. 2007 *Plant systematics: a phylogenetic approach*. SinauerAssoc, Sunderland.

[40] Kabiruddin M, and Makhzan-ul-Mufradat. 2007. New Delhi: Idara Kitabus Shifa; 314–315.

[41] Kadereit J.W. 2007. "Asterales: introduction and conspectus." In *Flowering Plants Eudicots*, vol 8, Springer, 1–6. Berlin.

[42] Kesseli R, Ochoa O, and Michelmore R. 1991. "Variation at RFLP loci in *Lactuca* spp. and origin of cultivated lettuce (*L. sativa*)." *Genome* 34: 430–436.

[43] Kim K.H, Kim Y.H, and Lee K.R. 2007. "Isolation of quinic acid derivatives and flavonoids from the aerial parts of *Lactucaindica* L. and their hepatoprotective activity in vitro." *Bioorganic & medicinal chemistry letters* 17: 6739–6743.

[44] Koopman W.J.M, Guetta E, van De Wiel C.C.M, Vosman B, and van den Berg R.G. 1998. "Phylogenetic relationships among *Lactuca* (Asteraceae) species and related genera based on ITS-1 DNA sequences." *American Journal of Botany* 85: 1517–1530.

[45] Koopman W.J.M, Zevenbergen M.J, van den Berg R.G. 2001. "Species relationships in *Lactuca*s.l. (Lactuceae, Asteraceae) inferred from AFLP fingerprints." *American Journal of Botany* 10: 1881–1887.

[46] Křístková E, Doležalová I, Lebeda A, Vinter V. and Novotná A. 2008. "Description of morphological characters of lettuce (*Lactuca sativa* L.) genetic resources." *Horticultural Science* 35: 113–129.

[47] Lebeda A, Dolezalová I, Feráková V, and Astley D. 2004. "Geographical distribution of wild *Lactuca* species (Asteraceae, Lactuceae)". *The Botanical Review* 70: 328–356.

[48] Lebeda A, Doležalová I, Křístková E, Kitner M, Petrželová I, Mieslerová B. and Novotná A. 2009. "Wild *Lactuca* germplasm for lettuce breeding: current status, gaps and challenges." *Euphytica* 170: 15–34.

[49] Lebeda A, Doležalová I, Křístková E. and Mieslerová B. 2001. "Biodiversity and ecogeography of wild *Lactuca* spp. in some

European countries." *Genetic Resources and Crop Evolution* 48: 153–164.

[50] Lebeda A, Ryder E.J, Grube R, Dolezalova I. and Kristkova E. 2007. "Lettuce (Asteraceae; *Lactuca* spp.)." *Genetic resources, chromosome engineering, and crop improvement series* 3: 377–472.

[51] Lebeda A. 1998. "Biodiversity of the interactions between germplasms of wild *Lactuca* spp. and related genera and lettuce downy mildew (*Bremialactucae*)." *Report on research programme OECD Biological Resource Management for Sustainable Agricultural Systems*, pp 70. HRI Wellesbourne (UK).

[52] Lebeda, A. and Astley, D. 1999. "World genetic resources of *Lactuca* spp., their taxonomy and biodiversity." In *Lebeda A. and Křístková E. (eds), Eucarpia Leafy Vegetables* 99. Palacký University, 81–94. Czech Republic.

[53] Lindqvist K. 1960. "On the origin of cultivated lettuce." *Hereditas* 46: 319–350.

[54] López A, Javier G.A, Fenoll J, Hellín P. and Flores P. 2014. "Chemical composition and antioxidant capacity of lettuce: Comparative study of regular-sized (Romaine) and baby-sized (Little Gem and Mini Romaine) types". *Journal of Food Composition and Analysis* 33: 39–48.

[55] Lopez E.G., and Jimenez A.C. 1974. *Elenco de la FloraVascular Espanola* (Península y Baleares). ICONA, Madrid (Spain).

[56] Lüthje P, Dzung D.N, Brauner A. 2011. "*Lactucaindica* extract interferes with uroepithelial infection by *Escherichia coli*." *Journal of Ethnopharmacology* 135: 672–677.

[57] Mampholo "B.M, Maboko M.M, Soundy P. and Sivakumar D. 2016. "Phytochemicals and overall quality of leafy lettuce (*Lactuca sativa* L.) varieties grown in closed hydroponic system." *Journal of food quality* 39: 805–815.

[58] Matvieieva N.A, Kudryavets Y.I, Likhova A.A, Shakhovskij A.M, Bezdenezhnykh N.A. and Kvasko E.Y. 2012. "Antiviral activity of extracts of transgenic chicory and lettuce plants with the human interferon α2b gene." *Cytology and Genetics* 46: 285–290."

[59] Michalska K, Stojakowska A, Malarz J, Doležalová I, Lebeda A. and Kisiel W. 2009. “Systematic implications of sesquiterpene lactones in *Lactuca* species.” *Biochemical Systematics and Ecology* 37: 174–179.

[60] Mou B. 2008. Lettuce. In *Prohens, J., Nuez, F. (Eds.), Handbook of Plant Breeding, Vol. I: Vegetables I: Asteraceae, -Brassicaceae, Chenopodicaceae, and Cucurbitaceae*. Springer, 75–116. New York, NY, USA

[61] Nessler C.L. 1976. “A systematic survey of the tribe Cichorieae in Virginia USA.” *Castanea* 41: 226–248.

[62] Niederwieser J.G. 2001. *Guide to Hydroponic Vegetable Production*, 2nd Ed., p. 140, Agricultural research council, roodeplaat vegetable and ornamental plant institute, Pretoria, South Africa.

[63] Oh M.M, Carey E.E. and Rajashekar C.B. 2011. “Antioxidant phytochemicals in lettuce grown in high tunnels and open field.” *Horticulture, Environment, and Biotechnology* 52: 133–139.

[64] Pascale S.D, Maggio A, Fogliano V, Ambrosino P. and Ritieni A. 2001. “Irrigation with saline water improves carotenoids content and antioxidant activity of tomato.” *The Journal of Horticultural Science and Biotechnology* 76: 447–453.

[65] Pepe G, Sommella E, Manfra M, De Nisco M, Tenore G.C, Scopa A, Sofo A, Marzocco S, Adesso S, Novellino T. and Campiglia P. 2015. “Evaluation of anti-inflammatory activity and fast UHPLC–DAD–IT-TOF profiling of polyphenolic compounds extracted from green lettuce (*Lactuca sativa* L.; var. *Maravilla de Verano*).” *Food chemistry* 167: 153–161.

[66] Qin X.X, Zhang M.Y, Han Y.Y, Hao J.H, Liu C.J. and Fan S.X. 2018. “Beneficial phytochemicals with anti-tumor potential revealed through metabolic profiling of new red pigmented lettuces (*Lactuca sativa* L.).” *International journal of molecular sciences* 19(4), p.1165.

[67] Rodenburg, C.M. 1960. *Varieties of Lettuce*: *An International Monograph*. TjeenkWillink, Zwolle W.E.J.

[68] Ruiz-Lozano J.M, Azcón R. and Palma J.M. 1996. "Superoxide dismutase activity in arbuscular mycorrhizal *Lactuca sativa* plants subjected to drought stress." *New Phytologist* 134: 327−333.

[69] Rulkens, A.J.H. 1987. *De CGN sla-collectie: inventarisatie, paspoortgegevens en enkele richtlijnen voor de toekomst.* Centre for Genetic Resources.

[70] Ryder E.J, Whitaker T.W. 1976. "Lettuce *Lactucasatira* (Compositae)." In *Simmonds Evolution of crop plants.* NW (ed) Longman, 39−41. London.

[71] Ryder, E.J. 1986. "Lettuce breeding." *Breeding Vegetable Crops*, pp.433−474.

[72] Sammour R.H. 2014. "Biochemical evaluation of *Lactuca* L. germplasm." *Research & Reviews in Bio Sciences Review* 8: 78−84.

[73] Sayyah M, Hadidi N. and Kamalinejad M. 2004. "Analgesic and anti-inflammatory activity of *Lactuca sativa* seed extract in rats." *Journal of Ethnopharmacology* 92: 325−329.

[74] Sharma S, Sheehy T, Kolonel L. 2014. "Sources of vegetables, fruits and vitamins A, C and E among five ethnic groups: results from a multiethnic cohort study." *The European Journal of Clinical Nutrition* 68: 384−391.

[75] Stebbins G.L. 1937. "The scandent species of *Prenanthes* and *Lactuca* in Africa." *Bulletin du Jardinbotanique de l'Etat a Bruxelles* 14: 333−352.

[76] Stojakowska A, Michalska K, Malarz J, Beharav A. and Kisiel W. 2013. "Root tubers of *Lactuca tuberosa* as a source of antioxidant phenolic compounds and new furofuranlignans." *Food chemistry* 138: 1250−1255.

[77] *The International Plant Names Index and World Checklist of Selected Plant Families 2020.* Published on the Internet at http://www.ipni.org and http://apps.kew.org/wcsp/.

[78] Urmila G.H, Rao B.G. and Satyanarayana T. 2013. "Phytochemical and in-vitro antioxidant activity of methanolic extract of *Lactuca scariola & Celosia argentea* leaves." *Journal of Drug Delivery and Therapeutics* 3: 114−117.

[79] USDA 2014. *National Statistics for Lettuce*. USDA, Washington, D.C.

[80] USDA. 2015. *National Nutrient Database for Standard Reference Release 28*. USDA, Washington, D.C.

[81] Van Treuren R, van der Arend A.J.M, Schut J.W. 2011. "Distribution of downy mildew (*Bremialactucae* Regel) resistances in a genebank collection of lettuce and its wild relatives. *Plant Genetic Resources* 11:15–25.

[82] Wang S.Y, Chang H.N, Lin K.T, Lo C.P, Yang N.S. and Shyur L.F. 2003. "Antioxidant properties and phytochemical characteristics of extracts from *Lactuca indica*." *Journal of agricultural and food chemistry* 51: 1506–1512.

[83] Wang Z.H, Peng H, Kilian N. 2013. "Molecular phylogeny of the *Lactuca* alliance (*Cichorieae subtribe* Lactucinae, Asteraceae) with focus on their Chinese centre of diversity detects potential events of reticulation and chloroplast capture." *PLoS ONE* 8: e82692.

[84] Wei Z, Zhu S.X, Van den Berg, R.G, Bakker F.T. and Schranz M.E. 2017. "Phylogenetic relationships within *Lactuca* L. (Asteraceae), including African species, based on chloroplast DNA sequence comparisons." *Genetic resources and crop evolution* 64: 55–71.

[85] Yadava R.N. and Jharbade J. 2007. "A new bioactive triterpenoid saponin from the seeds of *Lactucascariola* Linn." *Natural product research* 21: 500–506.

[86] Zdravkovic J, Pavlovic N, Pavlovic R, Maskovic P, Mladenovic J, Duric M. and Acamovic-Dokovic G. 2012. "Antimicrobial activity of lettuce (*Lactuca sativa* L.) extract grown in plastic and glasshouses." *Actahorticulturae*.

[87] Zekkori B, Khallouki F, Bentayeb A, Fiorito S, Preziuso F, Taddeo V.A, Epifano F. and Genovese S. 2018. "A new phytochemical and anti-oxidant and anti-inflammatory activities of different *Lactuca sativa* L. var. *crispa* extracts. *Natural Product Communications* 13: 1934578X1801300910.

[88] Złotek U. and Świeca M. 2016. "Elicitation effect of *Saccharomyces cerevisiae* yeast extract on main health-promoting compounds and

antioxidant and anti-inflammatory potential of butter lettuce (*Lactuca sativa* L.). *Journal of the Science of Food and Agriculture* 96: 2565−2572.

[89] Zohary D. 1991. "The wild genetic resources of cultivated lettuce (*Lactucasativa* L.)." *Euphytica* 53:31–35.

In: Lactuca: Cultivation and Uses
Editor: Jan Krüger
ISBN: 978-1-53617-729-9


*Chapter 2*

# *LACTUCA SATIVA*: AN EDIBLE SOURCE OF MEDICINAL BENEFITS

***Hammad Ismail*[1*]*, Sara Latif*[2]*, Fatima Ijaz*[2]
***and Erum Dilshad*[3]

[1]Department of Biochemistry and Biotechnology,
University University of Gujrat, Gujrat, Pakistan
[2]Department of Biochemistry, Quaid-i-Azam University,
Islamabad, Pakistan
[3]Department of Bioinformatics and Biosciences,
Faculty of Health and Life Sciences,
Capital University of Science and Technology (CUST),
Islamabad, Pakistan

## ABSTRACT

*Lactuca sativa,* commonly known as lettuce, belongs to the Asteraceae family. It is used as both delicious vegetable and important folk medicine. It is low in carbohydrate and fat contents with high water contents. Lettuce is rich with minerals and vitamins including vitamin A,

[*] Corresponding Author's E-mail: hammad.ismail@uog.edu.pk.

C, E, iron, potassium, calcium and phosphorus. It also contains selenium which has medicinal properties for the prevention of colon, prostate and lung cancers. This chapter emphasizes the medicinal properties of lettuce and its active components for different biological actions, for example, anti-inflammatory, antioxidant, antidiabetic, antimicrobial, analgesic, sedative and neuroprotective effects. The therapeutic potential of this plant is primarily due to sesquiterpene lactones which are considered as the most active compounds of lettuce. These compounds impart health benefits by providing a balanced diet and by acting as therapeutic agents, due to their strong curative potential against heart and cancer diseases. Lettuce is also a rich source of phytochemicals and plant secondary metabolites. These phytochemicals in lettuce have been reported for analgesic, hypnotic, sedative and anticonvulsant properties. Dietary phytochemicals are of specific interest due to their significance as antioxidants. Later in the chapter, we have discussed different techniques that have been applied to identify phenolic and phytochemicals compounds in lettuce like liquid chromatography and mass spectrometry. A detailed primary and secondary metabolite profiling of lettuce has also been described. Because lettuce has strong medicinal potential, its utilization in dietary supplementation can be further investigated. Moreover, clinical studies are also needed to assess the efficiency and safety of lettuce preparations and its isolated compounds for the therapeutic application.

**Keywords**: anticancer, antidiabetic, antioxidant, anti-inflammatory, metabolomics, neuroprotective

## INTRODUCTION

Lettuce is a cool-season leafy vegetable grown in approximately all continents of the world. It is mainly consumed as a fresh, raw salad vegetable [1] and some forms are also cooked. The genus *Lactuca* has been reported to have about three hundred species [2]. Earlier, Lindqvist [4] and colleagues have recognized about 100 species and listed about 70 species under the genus *Lactuca*. *Lactuca sativa L.* (lettuce) is a well-known edible crop that belongs to the *Lactuca* genus and the family Asteraceae. *Lactuca* means 'milk forming' and *sativa* means common [5]. Salad, tshilai and lettuce are the common names used for *Lactuca sativa*. It has ornamental and commercial importance so it is grown at a small scale in home gardens and large scale production is carried out all around the world for economic purposes [6]. It is a self-pollinated annual plant with 2n = 18 chromosomes [6]. It can be grown both under protected cultivation as well as in the open field [7].

## BRIEF HISTORY

Lettuce originated in the Mediterranean region and may have begun cultivation in Egypt as early as 4500 years BC [4]. Lettuces were supposedly grown by Persians 500 years BC and were brought into China during 600-900 AD. In 1494, Columbus brought into the New World a form of non-heading lettuce. This type quickly formed a seed stalk and in fact did not become a staple food crop. Head lettuce in the United States was first reported in 1543. Lettuce was also popular with the ancient Romans and Greeks and it arrived in the United States during colonial days [3]. It has been assumed that cultivated lettuce has emerged from weeds of *Lactuca serriola* [3]. Its origin has been explained by different theories that comprise of direct origin from *Lactuca serriola*, descent from wild *Lactuca sativa* and hybridization of different varieties. The gene flow between *L. sativa* and *L. serriola* has no barriers and both species are closely related [4].

## COMPOSITION OF *LACTUCA SATIVA*

### Nutrient Composition

Lettuce is common as low-calorie food that consists of plenty of water and many nutrients. The nutritional value of lettuce is mainly based on the morphological type, color, and orientation of the leaves as mentioned in Table 1. Lettuce is hierarchical at the seventh position as compared with other vegetables and fruits in terms of nutrient-rich contents [2]. Subject to the species, lettuce is a rich source of vitamin K, vitamin C, vitamin A and calcium in dark green cultivars along with copper, and iron. Depending upon the variety of lettuce, different lettuce has different dietary fibers like carbohydrates, proteins and a little number of fats [3, 4]. About 95% of lettuce consists of water having varying amounts of iron, copper, sodium and phosphorus (Table 1). Two key cultivars of lettuce consisting of Leaf lettuce and Cos lettuce have eminent levels of calcium, ascorbic acid and vitamin A [2, 4-6].

**Table 1. Nutrient data of different cultivars of *Lactuca sativa***

| **Component (Unit/100g of Fresh weight)** | **Butter Head** | **Cos/ Romaine** | **Crisphead/ Iceberg** | **Green Leaf** | **Red Leaf** |
|---|---|---|---|---|---|
| Water (g) | 95.63 | 94.61 | 95.64 | 94.98 | 95.64 |
| Energy (kcal) | 13 | 17 | 14 | 15 | 16 |
| Protein (g) | 1.35 | 1.23 | 0.9 | 1.36 | 1.33 |
| Total lipid (g) | 0.22 | 0.3 | 0.14 | 0.15 | 0.22 |
| Carbohydrate (g) | 2.23 | 3.29 | 2.97 | 2.87 | 2.26 |
| Fiber(g) | 1.1 | 2.1 | 1.2 | 1.3 | 0.9 |
| Sugars (g) | 0.94 | 1.19 | 1.97 | 0.78 | 0.48 |
| Calcium, Ca (mg) | 35 | 33 | 18 | 36 | 33 |
| Iron, Fe (mg) | 1.24 | 0.97 | 0.41 | 0.86 | 1.2 |
| Magnesium, Mg (mg) | 13 | 14 | 7 | 13 | 12 |
| Phosphorus, P (mg) | 33 | 30 | 20 | 29 | 28 |
| Potassium, K (mg) | 238 | 247 | 141 | 194 | 187 |
| Sodium, Na (mg) | 5 | 8 | 10 | 28 | 25 |
| Zinc, Zn (mg) | 0.2 | 0.23 | 0.15 | 0.18 | 0.2 |
| Vitamin C (mg) | 3.7 | 4 | 2.8 | 9.2 | 3.7 |
| Thiamin (mg) | 0.057 | 0.072 | 0.041 | 0.07 | 0.064 |
| Riboflavin (mg) | 0.062 | 0.067 | 0.025 | 0.08 | 0.077 |
| Niacin (mg) | 0.357 | 0.313 | 0.123 | 0.375 | 0.321 |
| Vitamin B-6 (mg) | 0.082 | 0.074 | 0.042 | 0.09 | 0.1 |
| Folate, DFE (mg) | 73 | 136 | 29 | 38 | 36 |
| Vitamin A (μg) | 166 | 436 | 25 | 370 | 375 |
| Vitamin A (IU) | 3312 | 8710 | 502 | 7405 | 7492 |
| Vitamin E (mg) | 0.18 | 0.13 | 0.18 | 0.22 | 0.15 |
| Vitamin K (μg) | 102.3 | 102.5 | 24.1 | 126.3 | 140.3 |
| Saturated (g) | 0.029 | 0.039 | 0.018 | 0.02 | 0.017 |
| Monounsaturated (g) | 0.008 | 0.012 | 0.006 | 0.006 | 0.005 |
| Polyunsaturated (g) | 0.117 | 0.16 | 0.074 | 0.082 | 0.072 |

**Table 2. Carotenoids and phenolic contents in different cultivars of *Lactuca sativa***

| Compound | Butter Head | Cos/Romaine | Crisphead/Iceberg | Green Leaf | Red Leaf |
|---|---|---|---|---|---|
| Carotenoids (mg/100g of Fresh weight) | | | | | |
| β-carotene | 1.99 | 5.23 | 0.30 | 4.44 | 4.50 |
| Lutein | 8.58 | 5.43 | 0.71 | 4.60 | 3.00 |
| Zeaxanthin | 0.24 | ND | 0.27 | 0.19 | 0.20 |
| Neoxanthin | 0.21 | 3.95 | 0.29 | ND | ND |
| Violaxanthin | 0.44 | 5.05 | 0.56 | ND | ND |
| Antheraxanthin | 3.27 | ND | ND | ND | ND |
| *Lactuca*xanthin | ND | 1.11 | ND | ND | ND |
| Phenolics (mg/100g of Fresh weight) | | | | | |
| Phenolic acid | 66.0 | 45.0 | 20.0 | 40.0 | 190.0 |
| Flavonoids | 20.0 | 17.0 | 1.0 | 24.0 | 120.0 |
| Anthocyanins | 20.0 | 780.0 | ND | 100.0 | 20.0 |
| Total phenolic | 120.0 | 260.0 | 20.0 | 570.0 | 330.0 |

ND = Not determined.

## Carotenoids and Phenolic Contents

Carotenoids belong to the lipid-soluble family of pigments commonly present in green leafy vegetables [7]. The principal compounds found in lettuce are lutein and β-carotene. Nevertheless, carotenoids amount differs with verities as described in Table 2. While phenolic compounds are the main secondary metabolites of plants responsible for the defense mechanism [8]. Phenolic compounds affect the sensory, nutritional and medicinal effects of foods [9]. Owing to their composition, phenolic components can be divided into different subgroups as presented in Table 2. Total phenolic contents are liable for the higher scavenging activity of the lettuce especially for peroxyl radicals (reactive oxygen species; ROS), as compared with many other vegetables and fruits [10]. These phytochemicals along with other vitamins and secondary metabolites have a strong antioxidant effect which helps in defending the injury triggered by ROS induced stress which is well known to cause damage to membranes and cell components in plants and animals [11]. So, the availability of antioxidants provides plants with distinct advantages. Usually, phytochemicals with antioxidant effects are vitamins such as ascorbic acid, carotenoids, tocopherol or polyphenols [12]. Additionally, these phytochemicals perform the same function in humans and plants. Humans are usually exposed to different oxidative stresses resulting from metabolic processes in the body, which can sometimes cause oxidative damage [13]. This oxidative stress may lead to degenerative diseases such as cataracts, cardiovascular, cancer and Alzheimer's disease in humans [14].

## Secondary Metabolites

Numerous reports have described that daily consumption of fruits and vegetables rich in polyphenols, provides positive health effects in the prevention of diseases, such as cardiovascular diseases and some types of cancer [15]. Many phenols including hydroxycinnamic acids and flavonoids (Table 3), present in many vegetables, have been reported for their strong antioxidant potential [16, 17]. Phytochemicals include a vast array of compounds like terpenoids, phenolics, alkaloids and vitamins [12] which are reported for various biological functions in plants. Numerous reports reflect the detection of the lettuce content of secondary metabolites (Table 3), leading to multiple biological functions being explained for the plant. The major biological active molecules are sesquiterpene lactones. Quantitative studies showed that the roots of *lettuce* contain 8-deoxylactucin, crepidiaside B and dihydrolactucin. These compounds were not detected in leaves [18]. However, when leaves are inoculated with pathogenic *Pseudomonas cichorii,* the lettuce synthesized lettucenin A and costunolide which possessed antimicrobial effects [19]. *L. sativa* contains two eudesmane sesquiterpene lactones including 1b-O-β-d-glucopyranosyl-4a-hydroxyl-5a,6b,11bH-eudesma-12,6a-olide and 1b-hydroxyl-15-O-(*p*-methoxyphenylacetyl)-5a,6b and 11bH-eudesma-3-en-12,6aolide identified in var. anagustata [20]. The seeds investigation of lettuce reported the identification of flavonol glycoside including, *Lactuca*sativoside A, japonicin A, caffeic acid and isoquercitrin [21].

**Table 3. Secondary metabolites in *Lactuca sativa***

| Sesquiterpene lactones | Hydroxycinnamic acid | Essential oils |
|---|---|---|
| Germacranolide lactuside A | Phaseolic acid | α-pinene |
| Guaianolides lactucin | Caffeic acid | γ-cymene |
| Lactucopicrin | Chicoric acid | Thymol |
| 8-deoxylactucin | Isochlorogenic acid | Durenol |
| Crepidiaside B | Limonene minor | α-terpinene, |
| Lettucenin A | β-pinene | Thymol acetate |
| Costunolide | α-terpinolene | Caryophyllene |
| 9b-hydroxyl-4b,11b,13,15- | Linalool | Spathulenol |
| 10b,14-dihydroxyl-11bH-guaiane4(15)- | 4-terpineol | Camphene |
| 1bhydroxyl-5a,6bH-eudesman-3-ene- | α-terpineol | **Others** |
| Macrocliniside A | O-methylthymol, | Syringin |
| 11b,13-dihydrolactucin | L-alloaromadendrene | Phytol |
| Cichorioside B | Viridiflorene | Cumene hydroperoxide |
| 10b,14-dihydroxy-10(14),11b(13)- | | Ethacrynic acid |
| Sesquiterpene lactones | **Phenolic acids** | **Flavonoids** |
| 1b-O-β-d-glucopyranosyl-4 ahydroxyl- | Caffeoyltartaric acid | Quercetin-3-O-galactoside |
| 1b-hydroxyl-15-O-(p- | Chlorogenic acid | Quercetin 3-(6- |
| 4a-O-β-d--glucopyranosyl15-hydroxyl- | Dicaffeoyltartaric acid | quercetin 3-(6- |
| Anthocyanin | Dicaffeoylquinic acid | Quercetin 3-glucoside |
| Cyanidin 3-malonylglucoside | | Quercetin 3-glucuronide |
| | | Luteolin-7-O-glucuronide |

One other study reported nine compounds from *L. sativa* extract including anthocyanin (01), phenolic acids (03) and flavonoids (05) [22]. Similarly, Mai and Glomb [16] isolated 11 compounds from *L. sativa.* The complete analysis showed that the presence of the volatile compounds and essential oils in *L. sativa* depending on the state of the plant (dry or fresh) [5]. *L. sativa* is rich in antioxidants compounds such as quercetin, vitamin C, caffeic acid, carotenoids, phytols and polyphenols [23]. The key conjugates present in extracts of *L. sativa* are 8-sulfate and 15-oxalyl constituents of the lactucin, guaianolide sesquiterpene lactones, lactucopicrin and deoxylactucin [24]. Due to the antioxidant property of these compounds, lettuce has been described to inhibit diseases such as cancer [25]. Overall, these reports suggested that herbal formulation from

*L. sativa* might be used as a potential agent in search of novel antimicrobials or antioxidants candidates [17, 26, 27].

## MEDICINAL PROPERTIES

The research on regularly consumed foods with medicinal potential can be a promising approach to find potent drugs without side effects. Recently, the use of herbal medicine to improve human health and to treat different human diseases has been increased. In this chapter, we have summarized the scientific knowledge about medicinal uses of *Lactuca sativa*, its phytochemicals, how they are investigated and their mode of action to recover symptoms of the disease and reduce its occurrence.

### Ethnopharmacological and General Uses

Lettuce is recommended for the patients who suffer from a moderate level of anxiety or sleeping problems, as it contains a sleep-inducing substance in it. In ancient times its seed-oil was used in traditional medicine as a sleeping aid and also to relieve inflammation and pain [28]. It is recommended to take lettuce before meal for the patients suffering from gastric pain and other digestive disorders, as it helps in digestion. The decoction of lettuce leaves is useful to treat asthma and bronchial spasms [3]. It is reported that lettuce plant was identified for the cure of issues related to the stomach, to enhance the digestive system and improve hunger and decrease inflammation [4]. Later, it was identified that due to the presence of anti-inflammatory compounds triterpene lactones [29] in lettuce, it is used to reduce inflammation, enhance appetite and stimulate digestion [30]. Lettuce is highly recommended for the patients of diabetes as it contains very low carbohydrate content and a higher concentration of magnesium. Due to its high magnesium content, it minimizes the risk of diabetes (type 2) because a recent meta-analysis found that greater the intake of magnesium, the lesser will be the chances of diabetes (type 2)

[31]. Lettuce is also a good source of Ω-3 polyunsaturated fatty acid, α-linolenic acid [32]. Such fatty acid form phospholipids bilayer that is linked to insulin sensitivity within the skeletal muscle and control diabetes [33].

The seed-oil of *Lactuca sativa* has been described for sedative, anticonvulsant, analgesic, hypnotic and effect [28]. Flavonoids and phenolic compounds, present in lettuce, are known to exert scavenging capabilities, anti-carcinogenic, anxiolytic and anti-inflammatory effects [34]. Formerly, a large number of pharmacological experiments have been carried out to determine the therapeutic effects of the extracts of *L. sativa* which exhibited anticonvulsant, hypnotic, sedative and antioxidant effects [35]. Furthermore, anti-nociceptive and anti-inflammatory properties of seed extracts have also been reported [30]. Inflammation is also characterized by heat, pain and redness. Depression is a shared symptom in pain patients with inflammation and both are reported to have similar mechanisms [36]. It has been reported that certain antidepressant drugs also contribute to integral alleviation of pain [37]. The essential oils and the crude extracts of lettuce also possess properties of antioxidant, antifungal or antimicrobial against infectious diseases [26]. It was investigated that alcoholic and aqueous extracts of lettuce seeds have an effective antispermatogenic property and serum level of testosterone in mice was enhanced by the aqueous extract. It was suggested that lettuce seed was an effective contraceptive agent [38].

## Antimicrobial Effects

Food spoilage is one of the leading problems in the world and is caused by microbes. Several diseases are caused by eating food spoiled by microbes like *E. coli, S. aureus* and *L. monocytogenes*. So, it is very important to use some chemicals to conserve food from microbial spoilage during food processing. For this purpose, natural compounds that are obtained from plants have gained much importance. Continued advancement in the formation of the new antimicrobial products from a natural source such as a few species of vegetables play a vital role to overcome this issue. *L. sativa* is a vegetable crop that has gained more interest recently. *L. sativa* is rich in nutritional values, in past it was cultivated as a seasonal crop but now it is cultivated and utilized throughout the year. Because of different methods and applications for the cultivation of the *L. sativa*, fresh *L. sativa* is easily available throughout the year. Many studies revealed that *L. sativa* exhibits antimicrobial activities. Green synthesis of silver nanoparticles from the decoction of leaves of *L. sativa* was prepared. These silver nanoparticles were used to assess antimicrobial activity against a different strain of bacteria such as *B. subtilis*, *P. Aeruginosa*, *S. aureus* and a clinically obtained *K. pneumoniae*. In this study, the diameter of the inhibition zone of bacterial growth varied between 10 and 12 mm. The methanol extract of lettuce exhibited antimicrobial properties against clinically isolated pathogen such as *S. aureus, S. pyogenes, E. coli, B. subtilis and P. aeruginosa* and zones of inhibition were varied between 11.5-14 mm. Aqueous extract of *L. sativa* also exhibited antibacterial activity to prevent the growth of *B. subtilis and E. coli* [39]. In another study, methanol and aqueous extracts of *L. sativa* were used to inhibit the growth of Gram -ve and Gram +ve bacteria. The methanol extract showed antibacterial activity with minimal inhibitory concertation (MIC) of 2.5 mg/ml against all tested bacteria and MIC value of aqueous extract was 5 mg/ml [40]. In another study, the antifungal and antibacterial effect of the *L. sativa* extract was reported against different strains including *S. aureus*, *K. pneumoniae*, *E. coli*, *P. vulgaris*, *P. mirabilis*, *B. suptilis*, *C. albicans* and *A. niger*. Least susceptibility toward ethanolic extract of *L. sativa* was shown by the *S. aureus* and *P. vulgaris* (MIC = 78, 125 μg/ml) bacteria, and all microbes exhibited significant susceptibility (MIC = 39, 1 μg/ml) [41].

In a study, three extracts of *L. sativa* (methanol, petroleum ether and ethyl acetate water) were used to measure the antimicrobial activity. All extracts exhibited antimicrobial properties. The petroleum ether extract eliminates the growth of all strains and the zones of inhibition were between 14-20 mm. The ether extract inhibited the growth of *S. aureus, E. coli* and *Y. enterocolitica*, inhibition was 19 mm, 20 mm and 19 mm respectively. The water extract inhibited the growth of *G. candidum* and *B. cinerea,* inhibition was 20 mm and 12 mm, respectively [42]. *L. sativa* also exhibits antifungal property, it is reported that latex saps from *L. sativa* can damage the cell wall and also change the cytoplasmic material of the yeast [43]. Methanol and aqueous extract of the lettuce possesses antiviral property against human cytomegalovirus AD-169 and coxsackie B virus type 3 [40].

## Antianxiety Effects

One-eighth of the total world population is suffering from anxiety, and it is being considered a significant research topic in the domain of psychopharmacology in the present decade. Anxiety has been regarded as one of the supreme prevailing as well as the best treatable mental disorder. Behavioral, physical, cognitive and emotional components can altogether be seen in anxiety. Benzodiazepines establish the most important class of compounds being utilized and prescribed as an anxiety treatment. Nonetheless, safety concerns over a thin boundary between the anxiolytic effects of benzodiazepines and those of harmful after-effects have led to the development of new compounds with less unwanted side effects [44]. Plants have been utilized for pharmaceutical purposes since the old times, and they are still used in medicines at a large scale throughout the world [45]. As anxiolytics, various kinds of herbal medicines were involved. *Lactuca sativa*, traditionally known as lettuce, has certain medicinal properties. Anticonvulsant, anti-inflammatory, sedative and hypnotic effects have been reported for the lettuce due to the presence of numerous phytochemicals [29, 46].

Among the animal models used for anxiety, elevated plus maze is mostly used, settled on the basis of findings that rats circumvent open elevated alleys, along with the belief that the evasion is created due to fear. Using some physiological, behavioral and pharmacological parameters in 1985 Pellow and his colleagues endorsed the test [44]. A study was conducted in which lettuce extract had been shown to have the potential to lessen anxiety levels in animals. This study indicated that lettuce has a rich source of natural antioxidants, which is capable of decreasing anxiety and it might aid in the prevention of anxiety progression [46]. Yet, the underlying components causing anxiety were uncertain until a few years ago. Regarding this scenario, mice were orally administered with the dried lettuce leaves extract for fifteen or thirty days and the locomotor plus anxiolytic effect was determined. This study reported that the alcoholic extract of *L. sativa* leaves having a huge amount of polyphenols and secondary metabolites is an effective anxiolytic means [47]. In former studies, the HPLC study of polyphenols present in Lettuce showed the incidence of vanillin, caffeic acid, chlorogenic acid, epicatechin, rutin hydrate, rutin and quercetin. Anxiolytic activity of the extract may be promoted by these compounds [47]. Rutin has been reported to possess antianxiety properties [48, 49] with the possibility of affecting the GABA and serotonin systems. GABA is largely recognized as taking part in the progression of depression, therefore diazepam which is a GABA agonist, showing short-term effectiveness in alleviating anxiety.

In another study, *Lactuca sativa* improved the exploratory behavior in addition to the biochemical changes occurring during the anxiolytic condition. Using two treatment durations i.e., for 15days and for 30 days, animals were fed with lettuce and were subjected to a number of behavioral parameters such as elevated T maze, elevated plus maze, rat exposure test, marble burying trial, new object recognition test, open field test and hyponeophagia. Mice group treated with the lettuce plant extract was found to have improved exploratory behavior and locomotor activity [46, 50]. Results of various experimental studies demonstrated that acute administration of extracts of *L. sativa* has anxiolytic effects in rats. The outcomes also support that traditional measurements of the elevated plus-maze are reliable and sufficient for the detection of anxiolytic properties of *L. sativa*. Yet, more studies are suggested to demonstrate the active constituent along with the exact mechanisms underlying these effects.

## Hypnotic and Sedative Effects

One-third of the total life span of humans is spent in sleeping, and the quality and quantity of sleep relate to health, productivity and mood stability. Cognitive performance including memory and learning are also dependent on a sufficient amount of sleep [51, 52]. Sleep disorders have close relation with mental and physical disabilities; lethargy, tiredness, memory problems, etc. Insomnia has been reported to be involved in obesity and cardiovascular disorders [53]. During the past few years, several findings have shown that active components or the extracts of natural plants get attached to the receptor GABAA-BZD, resulting in the promotion of sleep [54, 55]. Multiple natural herbs have shown a positive role in sleep or sedation including hops (*Humulus lupulus*), St. John's wort (*Hypericum perforatum*), Valerian (*Valeriana officinalis*), lingzhi (*Ganoderma lucidum*), chamomile (*Matricaria chamomilla*), maca and barley grass in Asia, U.S.A. and the Europe [54, 56]. A mixture of terpenes, flavonoids and other substances present in these herbs influences brain function by targeting specific receptors such as serotonin, histamine, orexin, melatonin and $\gamma$-aminobutyric acid (GABA) receptors. These natural herbs have been used separately as well as in combinations to increase the efficacy [57]. Certain combinations of herbal extracts have been reported as better dietary management elements. In an earlier study, it was shown that leaves of two *Lactuca sativa* varieties; roman lettuce and skullcap lettuce played a role in sleep progression [57, 58]. It was found that romaine lettuce contained additional phenolic compounds (chicoric, chlorogenic and caffeic acid). Oxidative strain and inflammation are closely associated with a greater incidence of disturbed sleep, so these elements being antioxidants are likely to be indirectly affecting the sleep pattern [59]. This study showed that the mixture of Romaine lettuce extract and Skullcap extract (RE/SE) produces a resilient sleep-promoting effect as compared to the treatment of a single herb. The additional effect of RE/SE mixture shows that the sleep producing components being derived from two herbs are numerous. The RE/SE mixture enhances the quality plus quantity of sleep by interacting with GABAA receptors and thus can be used to treat insomnia and other sleep disorders [60].

Leaves of *Lactuca sativa* have been studied for sedative, anticonvulsant and hypnotic properties. In a spontaneous locomotor activity test, lactucopricin and lactucin showed the sedative property in rats [61]. Alcoholic extract of *L. sativa* showed soothing outcome in toads, resulting in reduced motor activity and behavior. *L. sativa* also possesses a tranquilizing outcome. This aids in calming our nerves to regulate palpitations and to have a sound sleep during nights, thus evading sleeplessness. Metabolite profiling of lactones obtained from lettuce varieties indicated the sedative property is grated by lactucin, which is one of the major components present in lettuce juice [4]. Hypnotic and sedative medicines can have harmful effects on pregnant women, in addition, they can also endanger the fetus along with mother [62]. Very limited studies examined the traditional and herbal medicine for their effects on sleeplessness in pregnancy time. A study was carried out for the investigation of the effects of seeds of lettuce on insomnia in pregnancy time. An initial study showed encouraging the hypnotic effect of oil from seeds of lettuce on sleeplessness. Seeds of lettuce could be utilized in a hypnotic treatment which is nontoxic for the pregnant women having no after-effects and appears to be ascribable to seeds of lettuce at a certain dosage. Seeds of lettuce is a worthy natural remedy to treat insomnia during pregnancy [63].

In another study, alcoholic extract of *Lactuca sativa* elongated pentobarbital-induced duration of sleep at 400 mg/kg BW. The butanol fraction of the same extract has also prolonged the time period of sleep with a reduction in sleep latency at intensities equivalent to diazepam. It was revealed that the extract itself and fractions of the extract are lacking toxic effect signifying that lettuce promotes pentobarbital hypnotic effect with the absence of key harmful results and the compounds being accountable for this effect are most possibly the agents nonpolar in nature present in butanol fraction [64]. Phytol, which is diterpenoid in nature and is extracted from the ethanolic fraction of lettuce showed sedative property on the body by increasing the amount of gamma-aminobutyric acid within the central nervous system and then hindering the role of one of the key enzymes carrying out the breakdown of the neurotransmitter, i.e., succinic semialdehyde dehydrogenase [5].

## Neuroprotective Effects

Management of ischemia through the discovery of some new natural products having potential antioxidant properties is an exciting approach nowadays. Various plants having antioxidant properties have been found to prevent neurons from ischemic insult [65, 66]. The ethanolic extract of *L. sativa* has displayed shielding effects against D-galactose-mediated oxidative damage in neurons. Additionally, through its antioxidant activity, lettuce has been reported to inhibit $H_2O_2$- induced neurotoxicity [67]. In a former study, the ethyl acetate portion of the hydroalcoholic extract of *L. sativa* showed significant cytoprotective properties [68]. Another study was conducted to further examine the underlying mechanisms responsible for neuron protection effect of this part against the cytotoxic effect induced by GSD within N2a plus PC12 cell lines. The GSD situation depicts an *in vitro* system stimulating neuronal impairment during ischemia [65, 66]. This study showed overall positive effects of lettuce in protection against ischemia-induced neurotoxicity. Results of comet assay showed that lettuce is helpful in the prevention of neuronal apoptosis and fragmentation of DNA. The anti-apoptotic result was evident in the two cell lines used, whether treated with high or low amounts of lettuce. The Western Blotting revealed that the antiapoptotic effect of *L. sativa* has been facilitated through the reduction of caspase-3 and Bax/Bcl-2 ratio. Okada and Okada [69] stated that due to scavenging activity and greater phenolic content, an aqueous extract of the lettuce seeds guards against cell death and Aβ-mediated oxidative stress in hippocampal neurons. Results of this study propose that *L. sativa* could be considered as a possible nutraceutical mediator having strong shielding activities against certain neurodegenerative diseases such as cerebral ischemia, Alzheimer's disease and vascular dementia [70].

Several reports depicting the neuroprotective effects of lettuce are available. A study of lettuce was carried out for explaining the shielding effects of ethanol extract on the buildup of lipofuscin granules in different brain parts of D galactose driven female albino mice [71]. *L. sativa* reduced the granules of lipofuscin in the hippocampus and cerebral cortex and this might be a result of the creation of advanced glycation end products. *L. sativa* is an enriched source of antioxidants and the antioxidants all together should have abolished the consequence of D-galactose. Decreased lipofuscin granules in the group being treated with *L. sativa* depict lysosomal clearance which is an indication of a healthy cell. Lettuce protected the neuronal cells from lipofuscin granules accumulation and hence proved as an effective neuroprotective agent [4]. Another study was carried out to find out if *Lactuca sativa* is able to protect neuroblastoma-like cell lines of mouse i.e., PC12 and N2a against glucose/serum deprivation (GSD) state, presenting an *in vitro* model simulating cell damage induced by ischemia [72-74]. The outcomes of this work depicted that the neuroprotective property of *L. sativa* against GSD is due to the intermediate antioxidants which are water-soluble such as flavonoids. It is also reported that the progression of inflammation is involved in neural damages due to ischemia and anti-inflammatory agents can be utilized in cerebral stroke [75]. As anti-inflammatory activities were reported for lettuce, so prevention of inflammation can be additional possible mechanisms for relieving effects of lettuce in ischemic insults [29].

Keeping in view the conventional use of lettuce as a memory enhancer, its neuroprotective effect and various other effects on the CNS [70], scientists carried out a study to further inspect the result of standard ethanolic extract of *Lactuca sativa* (LSEE) along with its divisions against amnesia prompted by scopolamine. Effect of LSEE on memory was evaluated through different behavioral assays such as Morris water maze, object recognition and elevated plus-maze tasks in addition to biochemical factors were evaluated through different oxidative stress parameters and acetylcholinesterase activity. This study revealed that lettuce has the potential for memory enhancement against scopolamine-induced memory deficiencies. LSEE along with its fractions enhanced cognition without any harm to spontaneous locomotor activity in animals being treated with scopolamine. In addition after treatment with lettuce extracts, animals showed remarkable progress in spatial as well as non-spatial memory being compromised through scopolamine [76]. In this study, LSEE extract showed the existence of flavonoids and being standardized on the availability of quercetin fraction. Previously quercetin was found to depict antioxidant and acetylcholinesterase inhibition activity. Quercetin displayed protective effects against neurodegenerative impairments such as Alzheimer's disease [77]. Moreover, quercetin was shown to decrease amyloid-beta protein levels [78]. Hence, the memory-improving effect of LSEE could be due to the presence of quercetin.

## Antioxidant Effects

Oxidative damage is caused by the imbalance between the production and accumulation of ROS the reactive oxygen species, which are mainly the byproducts of oxygen metabolism and the capacity of a biological system to utilize and detoxify the cell from these free radicals. [79]. According to the classical oxidative stress theory of disease, this imbalance is the root cause of the pathophysiology and pathogenesis of several disorders [80] including neurodegenerative disorders such as Parkinson, Alzheimer's, cardiovascular disorders [81]. *Lactuca sativa*, is known to have therapeutic significance. As per available folk literature, it is utilized for the cure of a number of diseases such as insomnia, acute dry cough, anxiety, and neurosis [4]. Evidence of antioxidant properties of *L. sativa* has been reported through numerous *in vitro* and *in vivo* studies. Harsha and Anilakumar [47] demonstrated the antioxidant potential of the hydro-alcohol, leaf decoction of *L. sativa* through the detailed examination of the antioxidant and biomolecular protective effect of the extract. This study showed that alcoholic extract of *L. sativa* had strongly scavenged the DPPH (2,2-diphenyl-2-picryl hydrazyl), superoxide and nitric oxide ions with the $IC_{50}$ values 259 μg/mL, 162.8 μg/mL and 492 μg/mL when checked at the different concentration. The alcoholic extract of *L. sativa* also reduced the levels of free ferric ions and ferrous ions when checked through FRAP assay and metal chelating power assay respectively. *L. sativa* leaf extract also successfully prevented the peroxidation of lipid moieties, oxidation of proteins and oxidative stress-induced damage of deoxyribonucleic acids (DNA).

Materska et al. [82] reported the antioxidant activity of polyphenols extracted from *Lactuca sativa*. The study showed the potent radical DPPH scavenging potential of polyphenolic compounds obtained from powdered lettuce leave samples, where the phenolic content was recorded as 10.75 mg/g and a linear correlation of the total phenolic content and potential for DPPH scavenging was identified. Assefa et al. [83] reported the antioxidant potential of phenolic acids and flavonoids isolated at different growth stages of *L. sativa*. They also checked the antioxidant activity by ABTS decolorization assay. This study concluded that the mature, full-grown plants of *L. sativa* showed an increase of 2.1% to 40.9% in ABTS scavenging activity as compared to other growth stages. Not only the fresh green leave extracts of *L. sativa* presents the antioxidant potential but its other components including essential oils also showed the significant free radical scavenging activity. Al-Nomaani et al. [26] extracted the essential oils and volatile components from fresh and dried leaves of *L. sativa*. This study indicates that at least five different crude oil extracts from *L. sativa* possess a strong proton donating capacity so act as strong natural antioxidants.

Another study [76] reported the antioxidant properties of standardized *Lactuca sativa* ethanol extract and its various fractions including hexane, butanol and ethyl acetate in the male Laca mice model. In this study, activity was determined by various biochemical parameters such as superoxide dismutase, malonaldehyde, catalase, reduced glutathione, acetylcholinesterase activity and nitrite assay. Ethanol extract (200 mg/kg) of *L. sativa* and its butanol fraction (15 mg/kg) significantly ameliorates the oxidative stress through its acetylcholinesterase inhibition activity and antioxidant activity. Adesso et al. [84] scrutinized the polyphenolic extract of *L. sativa* for its antioxidant potential at the molecular level by using murine monocyte macrophage-based cell line, J774A.1. The study indicated that *L. sativa* enhances the expression of the HO-1 enzyme, which is believed to produce several antioxidants including bilirubin that has the capacity to stop the expression of iNOS and hence stops the production of NO.

## Anti-Inflammatory Effects

Inflammation is a complex localized physical reaction in response to tissue damage, pathogen attack or allergens that include immune cells and molecular mediators. *Lactuca sativa* is a traditionally famous vegetable for the management of acute and chronic inflammation. Araruna and Carlos [29] demonstrated the anti-inflammatory activity of triterpene lactones isolated from *L. sativa*. The study indicated the anti-inflammatory potential of triterpene lactones of *L. sativa* by conducting the in-vitro lipoxygenase inhibition assay by using serial dilutions of isolated components. The compounds revealed promising inhibitory effects on lipoxygenase enzyme with an IC50 value 59.0 μM which was comparable to the positive control Baicalein (IC50 22.1 μM). In the same study, the efficacy of *L. sativa* derived triterpenes were checked against carrageenan-induced edema test in rats and significant inhibition of inflammation was recorded by comparing the results with control drug, indomethacin (5 mg/kg).

Ismail and Mirza [85] reported the anti-inflammatory capacity of *L. sativa* leave and seed extracts along with cell suspensions. This study revealed that the aqueous leaf extracts showed the highest anti-inflammatory activity. Zekkori et al. [86] checked the anti-inflammatory potential of hydro-alcohol extracts of *L. sativa*. The in-vitro anti-inflammatory potential was determined by the extent of protection against protein denaturation. The highest protective potential was shown by $H_2O$-EtOH (1:3) extract which had the highest polyphenols using acetylsalicylic acid as a positive control. Zekkori et al. [86] also demonstrated the in-vivo anti-inflammatory effect of the alcoholic extract by formalin-induced paw edema test at two different concentrations i.e., 200 mg/kg and 400 mg/kg and percentage inhibition in paw edema was 43.8% and 50.1% respectively. The results were much comparable to the positive drug Dexamethasone which showed the 49.7% inhibition in paw edema at the concentration of 10 mg/kg.

*Lactuca sativa* extracts strongly interact with the molecular components of the inflammatory cascade. Adesso et al. [84] reported a detailed study showing the effect of *L. sativa* extracts on the iNOC and COX-2 pathway. COX-2 is a well elucidated pro-inflammatory enzyme whose expression is kicked on by NO. This study indicated that *L. sativa* extracts strongly inhibit the expression of COX-2. Similarly, *L. sativa* extracts also reduce the level of TNF-α and IL-6 which are the inflammatory cytokines by inhibiting the pro-inflammatory transcription factor, P65 NF-κB translocation in J774A.1 macrophage cell line. When tested in mouse primary macrophage, *L. sativa* extract significantly reduced the ROS expression at all tested concentrations, these results obtained thus testify the anti-inflammatory and antioxidant potential of *L. sativa* in primary mouse macrophage.

## Analgesic Effects

Pain is the term known for the sensation of physical discomfort, caused by potential tissue damage. Pain can be classified into three distinct categories. The first is the nociceptive pain, a sensation in the result of noxious stimuli. The second is the inflammatory pain caused by the sensory sensitivity due to unavoidable tissue injury resulting in the activation of the inflammatory system and the third type of pain is term as pathological pain which is a non-protective, maladaptive and irreversible abnormal functioning of the nervous system. This is the clinical pain syndrome, a consequence of amplified sensory signals in the CNS with a low threshold pain [87]. The term analgesic is used for any drug acting to relieve or alley the sensation of pain. Analgesics are classified as anti-inflammatory analgesic drugs that reduce the local inflammation thereby relieving the pain and the other class is the opioids, the drugs directly acting on the brain, suppressing the stimulus of pain in CNS [88]. Oertel and Loetsch [88] mentioned in their study that most of the analgesics work as COX-2 inhibitors. Adesso et al. [84] reported the strong COX-2 inhibitory activity of *Lactuca sativa* with reference to its potent anti-inflammatory behavior. Therefor extracts of *L. sativa* can be utilized as an effective analgesic to relieve the pain. Ismail and Mirza [85] checked the analgesic activity of aqueous extract of seeds, leave and cell suspension of *L. sativa* by using hot plate analgesic tests on adult albino rats. The outcomes of this study revealed that all extracts of *L. sativa* had significant analgesic potential while leave extracts being the most effective among all. Aspirin was the positive control in this study. Younus et al. [89] recently reported the analgesic properties of in-vitro and field-grown Iceberg lettuce extracts. Methanol, ethyl acetate and aqueous extracts were prepared with the standard method. In this study, the highest activity (75.74%) was shown by the leave extracts of field-grown plants at the concentration of 200mg/kg followed by regenerated callus and regenerated shoot extracts that showed the activity 66.52% and 59.40% respectively. In another study, Ismail et al. [90] reported the enhanced analgesic activity of *L. sativa* transformed with *rol C* gene (an enhancer gene). The comparison of the analgesic potential of both transformed and untransformed plants by hot plate analgesic assay. The highest activity recorded in the experiment was 79.1% showed by transformed lettuce.

## Anticoagulant Effects

Anticoagulants are the drugs that prevent the undesired blood clots in blood vessels. The available anticoagulants are either vitamin K antagonists that function primarily as the blocking agents of vitamin K epoxide reductase thereby halting the formation of vitamin K dependant blood clotting factors or Heparins that are the peptides that bind and enhance the activity of antithrombin III [91]. *Lactuca sativa* is broadly known for its pharmacological activities including its anticoagulant potential. Ismail and Mirza [85] reported the anticoagulant potential of

*Lactuca sativa* cv. Grand Rapids aqueous extracts determined by the capillary tube method. The study indicated that due to the presence of potent antioxidant constituents, *L. sativa* extracts showed the anticoagulant activity comparable to that of the standard synthetic anticoagulant i.e., Aspirin. In this study, it was observed that the mean clotting time for aspirin and methanol; chloroform leaf extract was 122s and 110s respectively which showed the promising potential of *L. sativa* extract as an anticoagulant agent. Noumedem et al. [5] in an elucidated review also mentioned the anticoagulant property of *L. sativa*. In a recent study, Ismail et al. [92] documented the anticoagulant activity of cell suspension cultures of *L. sativa* by the capillary tube method. Aspirin at the concentration of 10mg/kg was designated as the positive control in the study. The study reveals a very promising anticoagulant effect of all the tested plant lines.

## Antidiabetic Effects

Diabetes is a group of metabolic and chronic disorders characterized by hyperglycemia which leads to the formation of defects in insulin secretion and insulin action [93]. Diabetes is prevalent worldwide as a chronic disease, with estimated digits of 371 million for adults living with diabetes, having a prevalence of 8.3%. Treatment of diabetes avoiding the side effects is still a difficult task [94]. In old days medicinal plants were used to control diabetes and this practice in many civilizations had developed a quest for the investigation of the anti-diabetic drugs from the medicinal plants used to cure diabetes through natural resources. Some of these plants are edible foods that serve the purpose of both nutrition and medication. *Lactuca sativa* is one of those plants which serves for both purposes. *L. sativa* leaves are the warehouses for the phytonutrients which are mostly carotenoids like lutein (β,ε-carotene-3,3'-diol) β-carotene and many more [95]. *L. sativa* shows the antidiabetic property, its decoction can reduce the absorption of glucose in the intestine and it also exhibits the hypoglycemic effect. *L. sativa* contains carotenoid named *Lactuca*xanthin which has inhibitory property against the α-amylase and α-glycosides [96].

*L. sativa* contains three main metabolites (2,3,5,4'-tetrahydroxystilbene 2-O-β-D glucoside (THSG), ferulic acid (FA) and 4-oxnonanoic acid (4-ONA) which can promote the secretion of GLP-1 and enhanced the inhibition of α-glucosidase and DPP-4 [97]. It is reported that the ethanolic extract of *L. sativa* possesses alpha-amylase and alpha-glucosidase inhibitory activity [98]. In another study, it is described that polyphenol-rich lettuce Rutgers Scarlet Lettuce (RSL) was produced by somaclonal variation in tissue culturing technique. RSL consumption can decrease hyperglycemia and increase insulin sensitivity. RSL is thus used as a functional food for the dietary control of diabetics [99]. In a former study conducted by our group, *in vitro* bioassays showed that *rol* genes (*rol* A and *rol* C) greatly enhanced the capability of lettuce to inhibit dipeptidyl peptidase-4, α-glucosidase and stimulate GLP-1 secretion. It is reported that *L. sativa* has 76 metabolites (phenolic, non-phenolic and sesquiterpene lactones compounds) few of them were changed by many thousand percents in RA (*rol* A) and RC (*rol* C). Ferulic acid levels were increased 3033–9777%, aminooxononanoic acid level was increased 1141–1803% and 2,3,5,4′tetrahydroxystilbene-2-O-β-D-glucoside level were increased 40,272–48,008%. Activities of the compound were confirmed by comparing commercially obtained standards. It is studied that *rol* gene transformation greatly changed the metabolome of lettuce and improved its antidiabetic properties [100].

## Anticancer Effects

Cancer is the most fatal disease affecting millions of people every year, worldwide. Approximately 9.6 million people died in 2018 all over the world due to different types of cancer, among which lung cancer and female breast cancer were the most common, followed by prostate cancer and colorectal cancer [101]. Breast cancer that accounts for 11.6% of the diagnosis is the second major diagnosed cancer [101]. In recent years, the health effects of natural products have gained attention due to their wider medicinal importance. In the latest era, the antioxidant potential of plant-based compounds has received considerable attention [102, 103]. It was studied that polysaccharides obtained from *L. sativa* were tested against liver (HEPG2), colon (HCT 116), breast (MCF7) and cervical (Hela) carcinoma cell lines. This study indicates that the polysaccharides inhibit concentration-dependent on the growth pattern of these carcinogenic cells [104]. *L. sativa* water extract was used to inhibit the production of HL-10 leukemia and MCF-7 breast cancer cells, it is related to the induction of the checkpoint kinase 2, activation of tumor suppressor p21 and also involved in downregulation of pro oncogene cyclin D1. The ethyl acetate fraction of *L. sativa* induced HL-60 cell death, which was associated with the acetylation of alpha-tubulin. The amount calculated for humans was, 3kg lettuce consumption can promote inhibition of 50% HL-60 proliferation [105]. Intake of the vegetables that were rich in phenolic compounds, were found effective to decrease the risk of having different kinds of cancer. It is reported that phenolic extracts *L. sativa* that was grown in nitrogen low environment show well anti-proliferative effects against CACO-2 cells, along with interfering in cell cycle and cause apoptosis than those lettuces that are grown in the high amount of nitrogen-containing environment [106]. Lettuce exhibits antitumor activity, it is effective in the case of the human hepatoma cell line Bel7402, human lung adenocarcinoma cell line A549, human cancer colorectal adenoma cell line HT-29 and HCT-8 human colon cancer cell line [107]. In another study, effective components in *L. sativa* which indicated a significant relation to decreasing colorectal cancer with β-carotene and ascorbic acid, whereas, Ca, vitamin E, and folate have no association to colorectal cancer [108].

## METABOLOMIC PROFILING

The availability of reliable data for the determination of the consumption of nutrients and other food components is still one of the major challenges [109]. In spite of having information about energy requirements and amounts of macronutrients, data on some compounds and nutrients is still unavailable. This is true for some trace elements, fatty acids, folate and micronutrients, as well as for bioactive compounds including phenolic compounds. The phenolic portion of food is very complex and, in spite of having been studied numerous phenolic compounds present in food still have not been characterized and many functions remain to be resolved [110]. The main reason behind this is the complexity and the chemical nature of phenolic compounds and their interaction with the other similar compounds. One of the main problems hampering rapid analyses of such compounds is the lack of appropriate purified standards e.g., secoiridoids and lignan compounds. Generally, phenolic compounds of natural origin are considered as weak acids having phenolic hydroxyl groups, tannins having slightly acidic nature and flavonoids [111]. The identification and biochemical analysis of such compounds by traditional methods are challenging due to the diversity and multiple reactions of compounds. Meanwhile, modern detection and isolation methods, such as liquid chromatography, LC coupled mass spectrometry (MS), tandem mass spectrometry, GC-MS [112-114] and NMR-based characterization provide a solution for this problem.

Metabolomics plays a major role in biological study as one of the novels "omics" approach. The strategy works with the detection of molecules involved in metabolism followed by their identification and quantitation by high throughput techniques. Currently, biotechnology expansion, makes it possible to obtain metabolomic data of plant extracts with highly advance techniques like LC-MS, which combines MS detection with chromatographic isolation and has earned acceptance for metabolomic analyses due to its accuracy, sensitivity and coverage of metabolome [115]. In addition, MS is a powerful tool for structural characterization and it requires good pre-separation due to the crucial complexity of plant extracts/fractions and the presence of several isomers of compounds causing problem in differentiation by MS. LC-MS signals deliver information about the m/z ratio and the retention time of the detected ions, they must be mapped to metabolites to understand the biochemical processes they describe [116]. A QTOF-MS (time of flight mass spectrometer) measures the time needed for charged ions of different masses to travel from source region to detector [117]. The result of an ion detector is calculated as a function of time to generate a range of the accurate isotopic pattern which reveals the empirical/molecular formula of a metabolite with a significant degree of reliability and mass accuracy [118]. Combining a quadrupole to a TOF analyzer in a mass spectrometer provides flexibility for collision experiments, selectivity, high resolving power, great sensitivity, accurate mass measurements and speed in scan mode. A hybrid QTOF (quadrupole time of flight) mass analyzer corresponds to the combination of a quadrupole mass filter with a TOF mass analyzer. The QTOF-MS can operate in different modes which are the TOF mode and the production scan mode. Different sources of ionization are available like ESI (electrospray ionization), especially works in the negative ion mode, which is a pretty good choice for analyzing small molecules [119]. Moreover, QTOF delivers selectivity by extracted ion chromatogram mode under overlapping peaks conditions. Thus, QTOF provides the detection and separation of small compounds that might co-elute and could be underestimated as compared with major ions, and are not being identified/detected by traditional methodologies. Liquid chromatography coupled to QTOF-MS has been used in targeted metabolic analysis but it is also suitable for the detailed profiling of plant metabolites [120].

Metabolomics enables investigation of an extensive range of metabolite instantaneously in a non-biased method [121]. As mentioned in Table 4 numerous techniques have been applied to identify lettuce metabolic profile in terms of phytochemicals, organic acids, amino acids and their derivatives by employing techniques such as LC-MS, GC-MS and NMR [100]. All these techniques are based on a comparison of m/z ratio at particular RT for specific analyte with reference to pure standards. For example, Lee et al. [113] identified nine different organic acids in lettuce using GC/MS analysis. A comparative HPLC analysis of 11 lettuce species was carried out on sesquiterpene lactones, publicized that *L. sativa* contained, germacranolide lactuside A, lactucin, guaianolides and lactucopicrin as major constituents [18]. Ismail et al. [100], Viacava et al. [122] and Abu-Reidah et al. [123] putatively identified 71, 115 and 171 compounds in different varieties of lettuce using UPLC-QTOF-MS correspondingly. These putative compounds include organic acids, amino acids, peptides, terpenoids, alkaloids, lipids and phenolic compounds. García et al. [124] and Garcia et al. [125] identified different metabolites in two lettuce cultivar including romaine and iceberg, they also quantified metabolic changes caused by enzymatic browning using UPLC-ESI-QTOF-MS. However, a detailed primary and [126] secondary metabolite profiling of lettuce was carried out by using RP-UPLC-ESI-QTOF-MS and -MS/MS [123]. The more detail of metabolites and their individual class is given in Table 4.

## Analysis of Raw Data

Metabolic profiling is a well-developed tool to categorize the degree of metabolic impact [127]. QTOF-MS analysis, together with multivariate analyses, permits to detect metabolic variations [128]. Simple raw data from the QTOF analysis of the lettuce extracts can be handled using different software as mentioned in Table 4. Data analysis software permits the precise evaluation of high-resolution spectral data by LC-MS. These software assists in the validation of chromatographic approaches and also helps in the identification of metabolites features through the spectral data. Using TransOmics the spectral data from LC-MS can be aligned to given pooled sample and then adduct ions can be deconvoluted followed by determination of ions abundance features at the threshold level calculated. After the analysis in TransOmics, the data is screened to eliminate variables containing more than 20% zero values which are measured as technical noise of the instrument. The data is then transferred to Simca or related software for multivariate analysis. For quality assurance, the data is scaled as center and analysis is performed using principal components analysis (PCA). If the samples are pooled in the form of a cluster in the center of scores plot, it represents the technical reproducibility of the methodology [129].

## Table 4. Metabolic profiling of *Lactuca sativa*

| Sr # | Name of technique | Cultivar | Identification Software | Identified Metabolites | Reference Database | References |
|---|---|---|---|---|---|---|
| 1 | UPLC-QTOF-MS | Grand Rapids | Progenesis software | Total: 76<br>Sesquiterpene lactones: 12, non-phenolic: 24, phenolic: 40 | KEGG, Phenol explorer, PubChem, ChemSpider, Metlin, HMDB, MassBank | [100] |
| 2 | GCxGC-TOF/MS + UPLC-IMS-QTOF/MS | Leaf and head lettuce | LECO Chroma TOF software | Total: 171<br>Amino acids: 17, carbohydrates: 21, lipids: 14, organic acids: 39, nucleotides: 05, polyphenols: 59, terpenoids: 08, others: 08 | KEGG, PlantCyc, NIST mass spectral database, Metlin, ResPect, Lipids databases | [131] |
| 3 | UHPLC-DAD-ESI-QTOF-MS | Red and green oak-leaf | MassLynx software | Phenolic compounds: 115<br>Novel: 48 | Spectral database | [122] |
| 4 | UPLC-ESI-QTOF | Romaine | Agilent Mass Hunter | Total: 15<br>Lysophospholipids: 04, oxylipin-Jasmon1ate metabolites: 06, phenolic metabolites: 06 | Metlin, CEBAS lettuce metabolites database | [124] |
| 5 | UHPLC-3QMS-MS | Iceberg, romaine, trocadero | Agilent Mass Hunter | Phenolic compounds: 25 | Spectral database | [132] |
| 6 | UPLC-ESI-QTOF-MS | Iceberg | Agilent Mass Hunter | Total: 22<br>Amino acids: 04, phenolic compounds: 05, sesquiterpene lactones: 05, fatty acids: 03, lysophospholipids: 03, others: 02 | Metlin, ChemSpider, ChEBI, KEGG, Lipid maps | [125] |
| 7 | UPLC-ESI-QTOF-MS | Baby, | Bruker data | Total: 171 | PubChem, | [123] |

| Sr # | Name of technique | Cultivar | Identification Software | Identified Metabolites | Reference Database | References |
|---|---|---|---|---|---|---|
| | | romaine, iceberg | analysis software | Organic acids: 11, amino acids/peptides: 42, nucleosides: 03, sesquiterpene lactones: 17, alkaloids: 02, iridioid: 01, phenolic compounds: 92 | ChemSpider, KEGG, SciFinder, Phenol-Explorer, Metlin, MassBank, KNApSAcK | |
| 8 | UHPLC-IT-TOF | Maravilla de Verano | Formula Predictor Shimadzu | Polyphenols and derivatives: 12 | Commercial standards | [133] |
| 9 | UHPLC | Romaine | AutoTune software | Phenolic compounds: 11 | Commercial standards | [114] |
| 10 | HPLC | British Hilde | Agilent software | Sesquiterpene lactones: 07 | Commercial standards | [18] |
| 11 | GC-MS | Chungchima | Formula Predictor Shimadzu | Organic acids: roots: 05, nutrient solutions: 07 | Commercial standards | [113] |
| 12 | NMR | Cortina | COSY, TOCSY, 1H–13C HSQC, 1H–13C HMBC, DOSY | Total: 58 Carbohydrates: 07, sugar alcohols choline: 02, phenols: 03, organic acids: 10, amino acids: 12, pheophytins: 02, carotenoids: 04, sterols: 02, lipids: 10, hydrocarbons: 01, unassigned: 05 | Standard Spectra | [126] |
| 13 | RP-HPLC/ESI-MS/MS | Grand Rapids | Calibration curves | Total: 15, plant hormones and their metabolites | Internal standards | [134] |

UPLC: Ultra Performance Liquid Chromatography, QTOF: Quadrupole Time-of-Flight, MS: Mass Spectrometry, GC: Gas Chromatography, UHPLC: Ultra-High-Performance Liquid Chromatography, DAD: Diode-Array Detection, ESI: Electrospray Ionization, IMS: Ion Mobility Spectrometry, NMR: Nuclear Magnetic Resonance, RP: Reverse Phase,

### Identification of Putative Ions

The putative identification of detected metabolites is generally carried out manually by using MassLynx software (Waters, USA) or other similar functioning software as described in Table 4. MassLynx has different functions for inspecting isotope patterns, elemental composition, theoretical monoisotopic patterns, nominal masses and average masses. The first step of identification is the determination of the elemental composition of each metabolite by creating a list of all possible molecular/empirical formulas based on a simple CHNO algorithm [100]. MassLynx provides diverse functionalities such as electron configuration, elemental range, equivalents of double-bonds, deviation error between the measured and theoretical mass in terms of ppm and Da. It also provides basic sigma properties by comparing isotope patterns of measured values of metabolites with theoretical ones. Generally, error values of 5 and 50 values for ppm and Da accordingly are accepted range for good analysis. Then the molecular formula, monoisotopic masses, chromatogram, after CHNO algorithm, MS spectra and their respective fragmentation pattern is compared with available lettuce literature and online databases such as Metlin, HMDB and MassBank which are used to identify the fragmentation patterns of putative metabolites. Other online databases such as Kegg Ligand Database, Phenol-Explorer, PubChem and ChemSpider as mentioned in Table 4 can also be consulted/cross-check to confirm metabolic pathways, general classification and their biological roles.

## Clinical Trials

It was reported that the antioxidant effects of the fresh *L. sativa* may be affected due to storage under modified atmosphere packaging. In this research, eleven healthy volunteers participated including 6 males and 5 females, who aged 29-45 and with body weights of 48.0-60.0 kg. Every individual ate 250g fresh *L. sativa* and their blood was taken at various time periods 0h (before consumption), 2h, 3h and 6h after consuming lettuce. Repeated the same experiment after 3 days on the same *L. sativa* and keep at 58°C under modified atmosphere packaging. After the first treatment of lettuce, the plasma total radical trapping antioxidant potential (TRAP) increased greatly after 6h than to the value measured under changed storage atmosphere packaging of *L. sativa*. Plasma TRAP, quercetin and p-coumaric enhanced from base values 2 and 3h after fresh *L. sativa* intake. Enhanced different blood parameter was also obtained such as caffeic acid at 3h, plasma β-carotene after 6h, vitamin C concentration after intake of fresh leave of lettuce and no variations were obtained after the consumption of the equal amount of leaves under modified-atmosphere packaging for all the measured markers [130]. It had been mentioned in another study that *L. sativa* seed oil has been useful in the treatment of sleep problems and has no risk and is effective in the treatment of wild to moderate anxiety and sleep problems [28]. It was studied on the double-blind randomized placebo-controlled trial of *L. sativa* seeds on pregnancy-related insomnia. In this clinical trial, 100 pregnant women were selected for those with sleeping disorders and women aged 20-45 years. They are divided into two groups, each with 50 participants. Participants of one group were assigned to take capsules having 1000 mg of *L. sativa* seeds and the second group received capsule having 1000 mg of placebo for two weeks. After two weeks, the effect of the *L. sativa* seeds and placebo capsules were examined on the quality and pattern of sleep by using the Pittsburgh Sleep Quality Index (PSQI) method. The result of this study indicated decreased insomnia during pregnancy in participants of the group receiving lettuce seed capsule and improvement in the sleeping pattern was observed as compared to other those were taking placebo capsule. Thus, it is observed that *L. sativa* seeds are a natural way to cure sleeping disorder during pregnancy [63].

## CONCLUSION

*Lactuca sativa* (lettuce) is the most important crop in the group of leafy vegetables cultivated worldwide. It seems to possess diverse medicinal effects due to the presence of several important nutrients, minerals, vitamins and sesquiterpene lactones. This plant has been used in ethnomedicine for the cure of many health disorders. Studies already have supported the plant as a source of antimicrobial, antioxidant, neuroprotective and anti-inflammatory agents. The main active ingredients are sesquiterpene lactones from this plant which have been shown to have beneficial effects on human health. Furthermore, the quantitative approach has proved that commercial lettuce is a good source of some natural antioxidants. In addition, the number of sesquiterpene lactones which are mainly reported for their anti-inflammatory, analgesic and sedative properties, suggesting their isolation and purification on an industrial scale. Conclusively, we may say that lettuce can be used in the manufacture of pharmaceutically relevant multipurpose products.

## REFERENCES


[1] El Said Said, D. (2012). Detection of parasites in commonly consumed raw vegetables. *Alexandria Journal of Medicine*, 48 (4): 345-352.

[2] Mou, B. (2008). Lettuce, In *Vegetables I*, pp 75-116, Springer.

[3] Kim, M. J. Moon, Y. Tou, J. C. Mou, B. and Waterland, N. L. (2016). Nutritional value, bioactive compounds and health benefits of lettuce (*Lactuca sativa* L.). *Journal of Food Composition and Analysis*, 49: 19-34.

[4] Anilakumar, K. Harsha, S. and Sharma, R. (2017). Lettuce: a promising leafy vegetable with functional properties. *Defence Life Science Journal*, 2 (2): 178-185.

[5] Noumedem, J. Djeussi, D. Hritcu, L. Mihasan, M. and Kuete, V. (2017). *Lactuca sativa*, In *Medicinal Spices and Vegetables from Africa*, pp 437-449, Elsevier.

[6] Ryder, E. J. (2012). *Leafy salad vegetables*, Springer Science and Business Media.

[7] Maiani, G. Periago Castón, M. J. Catasta, G. Toti, E. Cambrodón, I. G. Bysted, A. Granado-Lorencio, F. Olmedilla-Alonso, B. Knuthsen, P. and Valoti, M. (2009). Carotenoids: actual knowledge on food sources, intakes, stability and bioavailability and their protective role in humans. *Molecular nutrition & food research*, 53 (S2): S194-S218.

[8] Dai, J. and Mumper, R. J. (2010). Plant phenolics: extraction, analysis and their antioxidant and anticancer properties. *Molecules*, 15 (10): 7313-7352.

[9] Karakaya, S. (2004). Bioavailability of phenolic compounds. *Critical Reviews in Food Science and Nutrition*, 44 (6): 453-464.

[10] Caldwell, C. R. (2003). Alkylperoxyl radical scavenging activity of red leaf lettuce (*Lactuca sativa* L.) phenolics. *Journal of Agricultural and Food Chemistry*, 51 (16): 4589-4595.

[11] El-Beltagi, H. S. and Mohamed, H. I. (2013). Reactive oxygen species, lipid peroxidation and antioxidative defense mechanism. *Notulae Botanicae Horti Agrobotanici Cluj-Napoca*, 41 (1): 44-57.

[12] Blokhina, O. Virolainen, E. and Fagerstedt, K. V. (2003). Antioxidants, oxidative damage and oxygen deprivation stress: a review. *Annals of Botany*, 91 (2): 179-194.

[13] Boyer, J. and Liu, R. H. (2004). Apple phytochemicals and their health benefits. *Nutrition Journal*, 3 (1): 5.

[14] Ames, B. N. Shigenaga, M. K. and Hagen, T. M. (1993). Oxidants, antioxidants, and the degenerative diseases of aging. *Proceedings of the National Academy of Sciences*, 90 (17): 7915-7922.

[15] Jang, M. Cai, L. Udeani, G. O. Slowing, K. V. Thomas, C. F. Beecher, C. W. Fong, H. H. Farnsworth, N. R. Kinghorn, A. D. and Mehta, R. G. (1997). Cancer chemopreventive activity of resveratrol, a natural product derived from grapes. *Science*, 275 (5297): 218-220.

[16] Mai, F. and Glomb, M. A. (2013). Isolation of phenolic compounds from iceberg lettuce and impact on enzymatic browning. *Journal of Agricultural and Food Chemistry*, 61 (11): 2868-2874.

[17] Park, H. Cho, H. and Kong, K. (2005). Purification and biochemical properties of glutathione S-transferase from *Lactuca sativa*. *Journal of Biochemistry and Molecular Biology*, 38 (2): 232.

[18] Michalska, K. Stojakowska, A. Malarz, J. Doležalová, I. Lebeda, A. and Kisiel, W. (2009). Systematic implications of sesquiterpene lactones in *Lactuca* species. *Biochemical Systematics and Ecology*, 37 (3): 174-179.

[19] Takasugi, M. Okinaka, S. Katsui, N. Masamune, T. Shirata, A. and Ohuchi, M. (1985). Isolation and structure of lettucenin A, a novel guaianolide phytoalexin from *Lactuca* sativa var. capitata (Compositae). *Journal of the Chemical Society, Chemical Communications* (10): 621-622.

[20] Han, Y. F. Cao, G. X. and Xia, M. (2009). Two new eudesmane sesquiterpenes from *Lactuca sativa* var. anagustata L. *Chinese Chemical Letters*, 20 (10): 1211-1214.

[21] Xu, F. Zou, G. Liu, Y. and Aisa, H. (2012). Chemical constituents from seeds of *Lactuca sativa*. *Chemistry of Natural Compounds*, 48 (4): 574-576.

[22] Ferreres, F. Gil, M. I. Castaner, M. and Tomás-Barberán, F. A. (1997). Phenolic metabolites in red pigmented lettuce (*Lactuca sativa*). Changes with minimal processing and cold storage. *Journal of Agricultural and Food Chemistry*, 45 (11): 4249-4254.

[23] Kim, J. H. and Botella, J. (2004). Etr1-1 gene expression alters regeneration patterns in transgenic lettuce stimulating root formation. *Plant Cell, Tissue and Organ Culture*, 78 (1): 69-73.

[24] Sessa, R. A. Bennett, M. H. Lewis, M. J. Mansfield, J. W. and Beale, M. H. (2000). Metabolite profiling of sesquiterpene lactones from *Lactuca* species major latex components are novel oxalate and sulfate conjugates of lactucin and its derivatives. *Journal of Biological Chemistry*, 275 (35): 26877-26884.

[25] Koronowicz, A. A. Kopeć, A. Master, A. Smoleń, S. Piątkowska, E. Bieżanowska-Kopeć, R. Ledwożyw-Smoleń, I. Skoczylas, Ł. Rakoczy, R. and Leszczyńska, T. (2016). Transcriptome profiling of Caco-2 cancer cell line following treatment with extracts from

iodine-biofortified lettuce (*Lactuca sativa* L.). *PloS One*, 11 (1): e0147336.

[26] Al-Nomaani, R. S. S. Hossain, M. A. Weli, A. M. Al-Riyami, Q. and Al-Sabahi, J. N. (2013). Chemical composition of essential oils and *in vitro* antioxidant activity of fresh and dry leaves crude extracts of medicinal plant of *Lactuca sativa* L. native to Sultanate of Oman. *Asian Pacific Journal of Tropical Biomedicine*, 3 (5): 353-357.

[27] Bang, M. H. Choi, S. Y. Jang, T. O. Kim, S. K. Kwon, O. S. Kang, T. C. Won, M. Park, J. and Baek, N. L. (2002). Phytol, SSADH inhibitory diterpenoid of *Lactuca sativa*. *Archives of Pharmacal Research*, 25 (5): 643-646.

[28] Yakoot, M. Helmy, S. and Fawal, K. (2011). Pilot study of the efficacy and safety of lettuce seed oil in patients with sleep disorders. *International Journal of General Medicine*, 4: 451.

[29] Araruna, K. and Carlos, B. (2010). Anti-inflammatory activities of triterpene lactones from *Lactuca sativa*. *Phytopharmacology*, 1 (1): 1-6.

[30] Sayyah, M. Hadidi, N. and Kamalinejad, M. (2004). Analgesic and anti-inflammatory activity of *Lactuca sativa* seed extract in rats. *Journal of Ethnopharmacology*, 92 (2-3): 325-329.

[31] Larsson, S. and Wolk, A. (2007). Magnesium intake and risk of type 2 diabetes: a meta-analysis. *Journal of Internal Medicine*, 262 (2): 208-214.

[32] Liu, S. Serdula, M. Janket, S. J. Cook, N. R. Sesso, H. D. Willett, W. C. Manson, J. E. and Buring, J. E. (2004). A prospective study of fruit and vegetable intake and the risk of type 2 diabetes in women. *Diabetes Care*, 27 (12): 2993-2996.

[33] Ford, E. S. and Mokdad, A. H. (2001). Fruit and vegetable consumption and diabetes mellitus incidence among US adults. *Preventive Medicine*, 32 (1): 33-39.

[34] Acosta-Estrada, B. A. Gutiérrez-Uribe, J. A. and Serna-Saldívar, S. O. (2014). Bound phenolics in foods, a review. *Food Chemistry*, 152: 46-55.

[35] Hefnawy, H. T. M. and Ramadan, M. F. (2013). Protective effects of *Lactuca sativa* ethanolic extract on carbon tetrachloride induced oxidative damage in rats. *Asian Pacific Journal of Tropical Disease*, 3 (4): 277-285.

[36] Micó, J. A. Ardid, D. Berrocoso, E. and Eschalier, A. (2006). Antidepressants and pain. *Trends in Pharmacological Sciences*, 27 (7): 348-354.

[37] Krell, H. V. Leuchter, A. F. Cook, I. A. and Abrams, M. (2005). Evaluation of reboxetine, a noradrenergic antidepressant, for the treatment of fibromyalgia and chronic low back pain. *Psychosomatics*, 46 (5): 379-384.

[38] Ahangarpour, A. Oroojan, A. A. and Radan, M. (2014). Effect of aqueous and hydro-alcoholic extracts of lettuce (*Lactuca sativa*) seed on testosterone level and spermatogenesis in NMRI mice. *Iranian Journal of Reproductive Medicine*, 12 (1): 65.

[39] Nazri, N. M. Ahmat, N. Adnan, A. Mohamad, S. S. and Ruzaina, S. S. (2011). *In vitro* antibacterial and radical scavenging activities of Malaysian table salad. *African Journal of Biotechnology*, 10 (30): 5728-5735.

[40] Edziri, H. Smach, M. Ammar, S. Mahjoub, M. Mighri, Z. Aouni, M. and Mastouri, M. (2011). Antioxidant, antibacterial, and antiviral effects of *Lactuca sativa* extracts. *Industrial Crops and Products*, 34 (1): 1182-1185.

[41] Mladenovic, J. D. Acamovic-Djokovic, G. S. Pavlovic, R. M. Zdravkovic, J. Pavle, Z. M. and Milan, S. Z. *Antioxidant and Antimicrobial activities of lettuc*e.

[42] Rasheed, E. M. and Al-khazraji, S. M. (2019). Investigation of Antimicrobial Activities of *Lactuca* Sativa (L.) Extracts against Clinical Pathogens. *Al-Nisour Journal for Medical Sciences*, 1 (2): 302-310.

[43] Moulin-Traffort, J. Giordani, R. and Regli, P. (1990). Antifungal action of latex saps from *Lactuca sativa* L. and *Asclepias curassavica* L. *Mycoses*, 33 (7-8): 383-392.

[44] Komaki, A. Nasab, Z. K. Shahidi, S. Sarihi, A. Salehi, I. and Ghaderi, A. (2014). Anxiolytic effects of acute injection of hydro-alcoholic extract of lettuce in the elevated plus-maze task in rats. *Avicenna J Neuro Psych Physio*, 1 (1): e18695.

[45] Long, C. Li, S. Long, B. Shi, Y. and Liu, B. (2009). Medicinal plants used by the Yi ethnic group: a case study in central Yunnan. *Journal of Ethnobiology and Ethnomedicine*, 5 (1): 13.

[46] Harsha, S. N. and Anilakumar, K. R. (2013). Anxiolytic property of hydro-alcohol extract of *Lactuca sativa* and its effect on behavioral activities of mice. *Journal of Biomedical Research*, 27 (1): 37.

[47] Harsha, S. and Anilakumar, K. (2013). Anxiolytic property of *Lactuca sativa*, effect on anxiety behaviour induced by novel food and height. *Asian Pacific Journal of Tropical Medicine*, 6 (7): 532-536.

[48] Priprem, A. Watanatorn, J. Sutthiparinyanont, S. Phachonpai, W. and Muchimapura, S. (2008). Anxiety and cognitive effects of quercetin liposomes in rats. *Nanomedicine: Nanotechnology, Biology and Medicine*, 4 (1): 70-78.

[49] Bradley, B. Starkey, N. Brown, S. and Lea, R. (2007). Anxiolytic effects of *Lavandula angustifolia* odour on the Mongolian gerbil elevated plus maze. *Journal of Ethnopharmacology*, 111 (3): 517-525.

[50] Harsha, S. and Anilakumar, K. (2012). Effects of *Lactuca sativa* extract on exploratory behavior pattern, locomotor activity and anxiety in mice. *Asian Pacific Journal of Tropical Disease*, 2: S475-S479.

[51] Sarris, J. Panossian, A. Schweitzer, I. Stough, C. and Scholey, A. (2011). Herbal medicine for depression, anxiety and insomnia: a review of psychopharmacology and clinical evidence. *European Neuropsychopharmacology*, 21 (12): 841-860.

[52] Vorster, A. P. and Born, J. (2015). Sleep and memory in mammals, birds and invertebrates. *Neuroscience & Biobehavioral Reviews*, 50: 103-119.

[53] Åkerstedt, T. and Wright, K. P. (2009). Sleep loss and fatigue in shift work and shift work disorder. *Sleep Medicine Clinics*, 4 (2): 257-271.

[54] Shi, Y. Dong, J. W. Zhao, J. H. Tang, L. N. and Zhang, J.-J. (2014). Herbal insomnia medications that target GABAergic systems: a review of the psychopharmacological evidence. *Current Neuropharmacology*, 12 (3): 289-302.

[55] Chen, L. C. Chen, I. C. Wang, B. R. and Shao, C. H. (2009). Drug-use pattern of Chinese herbal medicines in insomnia: A 4-year survey in Taiwan. *Journal of Clinical Pharmacy and Therapeutics*, 34 (5): 555-560.

[56] Zeng, Y. Yang, J. Du, J. Pu, X. Yang, X. Yang, S. and Yang, T. (2014). Strategies of functional foods promote sleep in human being. *Current Signal Transduction Therapy*, 9 (3): 148-155.

[57] Jana, S. and Rastogi, H. (2017). Effects of caffeic acid and quercetin on *in vitro* permeability, metabolism and *in vivo* pharmacokinetics of melatonin in rats: potential for herb-drug interaction. *European Journal of Drug Metabolism and Pharmacokinetics*, 42 (5): 781-791.

[58] Xie, Z. Chen, F. Li, W. A. Geng, X. Li, C. Meng, X. Feng, Y. Liu, W. and Yu, F. (2017). A review of sleep disorders and melatonin. *Neurological Research*, 39 (6): 559-565.

[59] Kanagasabai, T. and Ardern, C. I. (2015). Contribution of inflammation, oxidative stress, and antioxidants to the relationship between sleep duration and cardiometabolic health. *Sleep*, 38 (12): 1905-1912.

[60] Hong, K. B. Han, S. H. Park, Y. Suh, H. J. and Choi, H. S. (2018). Romaine lettuce/skullcap mixture improves sleep behavior in vertebrate models. *Biological and Pharmaceutical Bulletin*, 41 (8): 1269-1276.

[61] Wesołowska, A. Nikiforuk, A. Michalska, K. Kisiel, W. and Chojnacka-Wójcik, E. (2006). Analgesic and sedative activities of lactucin and some lactucin-like guaianolides in mice. *Journal of Ethnopharmacology*, 107 (2): 254-258.

[62] Wang, L. H. Lin, H. C. Lin, C. C. Chen, Y. H. and Lin, H. C. (2010). Increased risk of adverse pregnancy outcomes in women receiving zolpidem during pregnancy. *Clinical Pharmacology & Therapeutics*, 88 (3): 369-374.

[63] Pour, Z. S. Hosseinkhani, A. Asadi, N. Shahraki, H. R. Vafaei, H. Kasraeian, M. Bazrafshan, K. and Faraji, A. (2018). Double-blind randomized placebo-controlled trial on efficacy and safety of *Lactuca sativa* L. seeds on pregnancy-related insomnia. *Journal of Ethnopharmacology*, 227: 176-180.

[64] Ghorbani, A. Rakhshandeh, H. and Sadeghnia, H. R. (2013). Potentiating effects of *Lactuca* sativa on pentobarbital-induced sleep. *Iranian journal of pharmaceutical research: IJPR*, 12 (2): 401.

[65] Ghorbani, A. Rakhshandeh, H. Asadpour, E. and Sadeghnia, H. R. (2011). Effects of *Coriandrum sativum* extracts on glucose/serum deprivation-induced neuronal cell death. *Avicenna Journal of Phytomedicine*, 2 (1): 4-9.

[66] Mousavi, S. Tayarani-Najaran, Z. Asghari, M. and Sadeghnia, H. (2010). Protective effect of *Nigella sativa* extract and thymoquinone on serum/glucose deprivation-induced PC12 cells death. *Cellular and Molecular Neurobiology*, 30 (4): 591-598

[67] Im, S. E. Yoon, H. Nam, T. G. Heo, H. J. Lee, C. Y. and Kim, D. O. (2010). Antineurodegenerative effect of phenolic extracts and caffeic acid derivatives in romaine lettuce on neuron-like PC-12 cells. *Journal of Medicinal Food*, 13 (4): 779-784.

[68] Sadeghnia, H. R. Farahm, S. K. Asadpour, E. Rakhsh, H. and Ghorbani, A. (2012). Neuroprotective effect of *Lactuca sativa* on glucose/serum deprivation-induced cell death. *African Journal of Pharmacy and Pharmacology*, 6 (33): 2464-2471.

[69] Okada, Y. and Okada, M. (2013). Protective effects of plant seed extracts against amyloid β-induced neurotoxicity in cultured hippocampal neurons. *Journal of pharmacy & bioallied sciences*, 5 (2): 141.

[70] Ghorbani, A. Sadeghnia, H. R. and Asadpour, E. (2015). Mechanism of protective effect of lettuce against glucose/serum deprivation-induced neurotoxicity. *Nutritional Neuroscience*, 18 (3): 103-109.

[71] Harsha, S. and Anilakumar, K. (2013). Protection against aluminium neurotoxicity: A repertoire of lettuce antioxidants. *Biomedicine & Aging Pathology*, 3 (4): 179-184.

[72] Mousavi, S. H. Tayarani, N. and Parsaee, H. (2010). Protective effect of saffron extract and crocin on reactive oxygen species-mediated high glucose-induced toxicity in PC12 cells. *Cellular and Molecular Neurobiology*, 30 (2): 185-191.

[73] Zhang, Q. Qian, Z. Pan, L. Li, H. and Zhu, H. (2011). The traditional Chinese medicine Huang-Lian-Jie-Du-Tang inhibits hypoxia-induced neuronal apoptosis. *African Journal of Pharmacy and Pharmacology*, 5 (23): 2558-2565.

[74] Liu, D. Wang, S. Zhang, L. and Li, L. (2012). Chemical composition of n-butanol extract of *Potentilla anserina* L. and its protective effect of EAhy926 endothelial cells under hypoxia. *African Journal of Pharmacy and Pharmacology*, 6 (10): 677-684.

[75] Iadecola, C. and Anrather, J. (2011). The immunology of stroke: from mechanisms to translation. *Nature Medicine*, 17 (7): 796.

[76] Malik, J. Kaur, J. and Choudhary, S. (2018). Standardized extract of *Lactuca sativa* Linn. and its fractions abrogates scopolamine-induced amnesia in mice: A possible cholinergic and antioxidant mechanism. *Nutritional Neuroscience*, 21 (5): 361-372.

[77] Maciel, R. M. Carvalho, F. B. Olabiyi, A. A. Schmatz, R. Gutierres, J. M. Stefanello, N. Zanini, D. Rosa, M. M. Andrade, C. M. and Rubin, M. A. (2016). Neuroprotective effects of quercetin on memory and anxiogenic-like behavior in diabetic rats: Role of ectonucleotidases and acetylcholinesterase activities. *Biomedicine & Pharmacotherapy*, 84: 559-568.

[78] Zhang, X. Hu, J. Zhong, L. Wang, N. Yang, L. Liu, C. C. Li, H. Wang, X. Zhou, Y. and Zhang, Y. (2016). Quercetin stabilizes apolipoprotein E and reduces brain Aβ levels in amyloid model mice. *Neuropharmacology*, 108: 179-192.

[79] Pizzino, G. Irrera, N. Cucinotta, M. Pallio, G. Mannino, F. Arcoraci, V. Squadrito, F. Altavilla, D. and Bitto, A. (2017). Oxidative stress: harms and benefits for human health. *Oxidative Medicine and Cellular Longevity*, 2017:

[80] Ghezzi, P. Jaquet, V. Marcucci, F. and Schmidt, H. H. (2017). The oxidative stress theory of disease: levels of evidence and epistemological aspects. *British Journal of Pharmacology*, 174 (12): 1784-1796.

[81] Poprac, P. Jomova, K. Simunkova, M. Kollar, V. Rhodes, C. J. and Valko, M. (2017). Targeting free radicals in oxidative stress-related human diseases. *Trends in Pharmacological Sciences*, 38 (7): 592-607.

[82] Materska, M. Olszówka, K. Chilczuk, B. Stochmal, A. Pecio, Ł. Pacholczyk-Sienicka, B. Piacente, S. Pizza, C. and Masullo, M. (2019). Polyphenolic profiles in lettuce (*Lactuca* sativa L.) after CaCl 2 treatment and cold storage. *European Food Research and Technology*, 245 (3): 733-744.

[83] Assefa, A. D. Choi, S. Lee, J. E. Sung, J. S. Hur, O. S. Ro, N. Y. Lee, H. S. Jang, S. W. and Rhee, J. H. (2019). Identification and quantification of selected metabolites in differently pigmented leaves of lettuce (*Lactuca sativa* L.) cultivars harvested at mature and bolting stages. *BMC Chemistry*, 13 (1): 56.

[84] Adesso, S. Pepe, G. Sommella, E. Manfra, M. Scopa, A. Sofo, A. Tenore, G. C. Russo, M. Di Gaudio, F. and Autore, G. (2016). Anti-inflammatory and antioxidant activity of polyphenolic extracts from *Lactuca sativa* (var. Maravilla de Verano) under different farming methods. *Journal of the Science of Food and Agriculture*, 96 (12): 4194-4206.

[85] Ismail, H. and Mirza, B. (2015). Evaluation of analgesic, anti-inflammatory, anti-depressant and anti-coagulant properties of *Lactuca sativa* (CV. Grand Rapids) plant tissues and cell suspension in rats. *BMC Complementary and Alternative Medicine*, 15 (1): 199.

[86] Zekkori, B. Khallouki, F. Bentayeb, A. Fiorito, S. Preziuso, F. Taddeo, V. A. Epifano, F. and Genovese, S. (2018). A new phytochemical and anti-oxidant and anti-inflammatory activities of different *Lactuca sativa* L. var. crispa extracts. *Natural Product Communications*, 13 (9): 1934578X1801300910.

[87] Woolf, C. J. (2010). What is this thing called pain? *The Journal of Clinical Investigation*, 120 (11): 3742-3744.

[88] Oertel, B. G. and Loetsch, J. (2013). Clinical pharmacology of analgesics assessed with human experimental pain models: bridging basic and clinical research. *British Journal of Pharmacology*, 168 (3): 534-553.

[89] Younus, I. Ismail, H. Rizvi, C. B. Dilshad, E. Saba, K. Mirza, B. and Tahir, M. (2019). Antioxidant, analgesic and anti-inflammatory activities of *in vitro* and field-grown Iceberg lettuce extracts. *Journal of Pharmacy & Pharmacognosy Research*, 7 (5): 343-355.

[90] Ismail, H. Dilshad, E. Waheed, M. T. Sajid, M. Kayani, W. K. and Mirza, B. (2016). Transformation of *Lactuca sativa* L. with rol C gene results in increased antioxidant potential and enhanced analgesic, anti-inflammatory and antidepressant activities *in vivo*. *3 Biotech*, 6 (2): 215.

[91] Harter, K. Levine, M. and Henderson, S. O. (2015). Anticoagulation drug therapy: a review. *Western Journal of Emergency Medicine*, 16 (1): 11.

[92] Ismail, H. Kayani, S. S. Kayani, S. I. Mirza, B. and Waheed, M. T. (2019). Optimization of cell suspension culture of transformed and untransformed lettuce for the enhanced production of secondary metabolites and their pharmaceutical evaluation. *3 Biotech*, 9 (9): 339.

[93] Association, A. D. (2013). Diagnosis and classification of diabetes mellitus. *Diabetes care*, 36 (Supplement 1): S67-S74.

[94] Ghazanfar, K. Ganai, B. A. Akbar, S. Mubashir, K. Dar, S. A. Dar, M. Y. and Tantry, M. A. (2014). Antidiabetic activity of Artemisia amygdalina Decne in streptozotocin induced diabetic rats. *BioMed research international*, 2014:

[95] Roman-Ramos, R. Flores-Saenz, J. and Alarcon-Aguilar, F. (1995). Anti-hyperglycemic effect of some edible plants. *Journal of Ethnopharmacology*, 48 (1): 25-32.

[96] Gopal, S. S. Lakshmi, M. J. Sharavana, G. Sathaiah, G. Sreerama, Y. N. and Baskaran, V. (2017). *Lactuca*xanthin–a potential anti-diabetic carotenoid from lettuce (*Lactuca* sativa) inhibits α-amylase and α-glucosidase activity *in vitro* and in diabetic rats. *Food & function*, 8 (3): 1124-1131.

[97] Ismail, H. Gillespie, A. L. Calderwood, D. Iqbal, H. Gallagher, C. Chevallier, O. P. Elliott, C. T. Pan, X. Mirza, B. and Green, B. D. The Health Promoting Bioactivities of *Lactuca* sativa can be Enhanced by Genetic Modulation of Plant Secondary Metabolites.

[98] Timothy, C. N. and Geetha, R. (2019). *In vitro* alpha-amylase and alpha-glucosidase inhibitory activity of the ethanolic extract of *Lactuca* sativa. *Drug Invention Today*, 11 (8):

[99] Cheng, D. M. Pogrebnyak, N. Kuhn, P. Krueger, C. G. Johnson, W. D. and Raskin, I. (2014). Development and phytochemical characterization of high polyphenol red lettuce with anti-diabetic properties. *PloS one*, 9 (3): e91571.

[100] Ismail, H. Gillespie, A. L. Calderwood, D. Iqbal, H. Gallagher, C. Chevallier, O. P. Elliott, C. T. Pan, X. Mirza, B. and Green, B. D. (2019). The health promoting bioactivities of *Lactuca sativa* can be enhanced by genetic modulation of plant secondary metabolites. *Metabolites*, 9 (5): 97.

[101] Bray, F. Ferlay, J. Soerjomataram, I. Siegel, R. L. Torre, L. A. and Jemal, A. (2018). Global cancer statistics 2018: GLOBOCAN estimates of incidence and mortality worldwide for 36 cancers in 185 countries. *CA: A Cancer Journal for Clinicians*, 68 (6): 394-424.

[102] Lam, M. Carmichael, A. R. and Griffiths, H. R. (2012). An aqueous extract of *Fagonia cretica* induces DNA damage, cell cycle arrest and apoptosis in breast cancer cells via FOXO3a and p53 expression. *PloS one*, 7 (6): e40152.

[103] Walker, H. (2011). Metabolic profiling of plant tissues by electrospray mass spectrometry. *Handbook of Molecular Microbial Ecology I: Metagenomics and Complementary Approaches*: 697-708.

[104] Moharib, S. A. El Maksoud, N. A. Ragab, H. M. and Shehata, M. (2014). Anticancer activities of mushroom polysaccharides on chemically-induced colorectal cancer in rats. *Journal of Applied Pharmaceutical Science*, 4 (7): 54.

[105] Gridling, M. Popescu, R. Kopp, B. Wagner, K.-H. Krenn, L. and Krupitza, G. (2010). Anti-leukaemic effects of two extract types of *Lactuca sativa* correlate with the activation of Chk2, induction of p21, downregulation of cyclin D1 and acetylation of α-tubulin. *Oncology Reports*, 23 (4): 1145-1151.

[106] Zhou, W. Liang, X. Dai, P. Chen, Y. Zhang, Y. Zhang, M. Lu, L. Jin, C. and Lin, X. (2019). Alteration of Phenolic Composition in Lettuce (*Lactuca* sativa L.) by Reducing Nitrogen Supply Enhances its Anti-Proliferative Effects on Colorectal Cancer Cells. *International journal of molecular sciences*, 20 (17): 4205.

[107] Qin, X. X. Zhang, M. Y. Han, Y. Y. Hao, J. H. Liu, C. J. and Fan, S. X. (2018). Beneficial phytochemicals with anti-tumor potential revealed through metabolic profiling of new red pigmented lettuces (*Lactuca sativa* L.). *International Journal of Molecular Sciences*, 19 (4): 1165.

[108] Fernandez, E. La Vecchia, C. D'Avanzo, B. Negri, E. and Franceschi, S. (1997). Risk factors for colorectal cancer in subjects with family history of the disease. *British Journal of Cancer*, 75 (9): 1381.

[109] Greenfield, H. and Southgate, D. A. (2003). *Food composition data: production, management, and use*, Food & Agriculture Org.

[110] Cheynier, V. (2012). Phenolic compounds: from plants to foods. *Phytochemistry Reviews*, 11 (2-3): 153-177.

[111] Yao, L. H. Jiang, Y. Shi, J. Tomas-Barberan, F. Datta, N. Singanusong, R. and Chen, S. (2004). Flavonoids in food and their health benefits. *Plant Foods for Human Nutrition*, 59 (3): 113-122.

[112] Cao, J. Chen, W. Zhang, Y. Zhang, Y. and Zhao, X. (2010). Content of selected flavonoids in 100 edible vegetables and fruits. *Food Science and Technology Research*, 16 (5): 395-402.

[113] Lee, J. G. Lee, B. Y. and Lee, H. J. (2006). Accumulation of phytotoxic organic acids in reused nutrient solution during hydroponic cultivation of lettuce (*Lactuca sativa* L.). *Scientia Horticulturae*, 110 (2): 119-128.

[114] Ribas-Agustí, A. Gratacós-Cubarsí, M. Sárraga, C. García-Regueiro, J. A. and Castellari, M. (2011). Analysis of eleven phenolic compounds including novel p-coumaroyl derivatives in lettuce (*Lactuca sativa* L.) by ultra-high-performance liquid chromatography with photodiode array and mass spectrometry detection. *Phytochemical Analysis*, 22 (6): 555-563.

[115] Dettmer, K. Aronov, P. A. and Hammock, B. D. (2007). Mass spectrometry-based metabolomics. *Mass Spectrometry Reviews*, 26 (1): 51-78.

[116] Huang, N. Siegel, M. M. Kruppa, G. H. and Laukien, F. H. (1999). Automation of a fourier transform ion cyclotron resonance mass spectrometer for acquisition, analysis, and e-mailing of high-resolution exact-mass electrospray ionization mass spectral data. *Journal of the American Society for Mass Spectrometry*, 10 (11): 1166-1173.

[117] Wenzel, R. J. Matter, U. Schultheis, L. and Zenobi, R. (2005). Analysis of megadalton ions using cryodetection MALDI time-of-flight mass spectrometry. *Analytical Chemistry*, 77 (14): 4329-4337.

[118] Hall, R. D. and Hardy, N. W. (2011). Practical applications of metabolomics in plant biology, In *Plant Metabolomics*, pp 1-10, Springer.

[119] Gómez-Romero, M. Segura-Carretero, A. and Fernández-Gutiérrez, A. (2010). Metabolite profiling and quantification of phenolic compounds in methanol extracts of tomato fruit. *Phytochemistry*, 71 (16): 1848-1864.

[120] Picó, Y. la Farré, M. Soler, C. and Barceló, D. (2007). Identification of unknown pesticides in fruits using ultra-performance liquid chromatography–quadrupole time-of-flight mass spectrometry: Imazalil as a case study of quantification. *Journal of Chromatography A*, 1176 (1-2): 123-134.

[121] Dunn, W. B. Bailey, N. J. and Johnson, H. E. (2005). Measuring the metabolome: current analytical technologies. *Analyst*, 130 (5): 606-625.

[122] Viacava, G. E. Roura, S. I. Berrueta, L. A. Iriondo, C. Gallo, B. and Alonso-Salces, R. M. (2017). Characterization of phenolic compounds in green and red oak-leaf lettuce cultivars by UHPLC-DAD-ESI-QTOF/MS using MSE scan mode. *Journal of Mass Spectrometry*, 52 (12): 873-902.

[123] Abu-Reidah, I. Contreras, M. Arráez-Román, D. Segura-Carretero, A. and Fernández-Gutiérrez, A. (2013). Reversed-phase ultra-high-performance liquid chromatography coupled to electrospray ionization-quadrupole-time-of-flight mass spectrometry as a powerful tool for metabolic profiling of vegetables: *Lactuca sativa* as an example of its application. *Journal of Chromatography A*, 1313: 212-227.

[124] García, C. J. García-Villalba, R. Gil, M. I. and Tomas-Barberan, F. A. (2017). LC-MS untargeted metabolomics to explain the signal metabolites inducing browning in fresh-cut lettuce. *Journal of Agricultural and Food Chemistry*, 65 (22): 4526-4535.

[125] Garcia, C. J. García-Villalba, R. Garrido, Y. Gil, M. I. and Tomás-Barberán, F. A. (2016). Untargeted metabolomics approach using UPLC-ESI-QTOF-MS to explore the metabolome of fresh-cut iceberg lettuce. *Metabolomics*, 12 (8): 138.

[126] Sobolev, A. P. Brosio, E. Gianferri, R. and Segre, A. L. (2005). Metabolic profile of lettuce leaves by high-field NMR spectra. *Magnetic Resonance in Chemistry*, 43 (8): 625-638.

[127] Dixon, R. A. Gang, D. R. Charlton, A. J. Fiehn, O. Kuiper, H. A. Reynolds, T. L. Tjeerdema, R. S. Jeffery, E. H. German, J. B. and Ridley, W. P. (2006). Applications of metabolomics in agriculture. *Journal of Agricultural and Food Chemistry*, 54 (24): 8984-8994.

[128] Ciborowski, M. Teul, J. Martin-Ventura, J. L. Egido, J. and Barbas, C. (2012). Metabolomics with LC-QTOF-MS permits the prediction of disease stage in aortic abdominal aneurysm based on plasma metabolic fingerprint. *PLoS One*, 7 (2): e31982.

[129] Graham, S. F. Chevallier, O. P. Roberts, D. Hölscher, C. Elliott, C. T. and Green, B. D. (2013). Investigation of the human brain metabolome to identify potential markers for early diagnosis and therapeutic targets of Alzheimer's disease. *Analytical Chemistry*, 85 (3): 1803-1811.

[130] Serafini, M. Bugianesi, R. Salucci, M. Azzini, E. Raguzzini, A. and Maiani, G. (2002). Effect of acute ingestion of fresh and stored lettuce (*Lactuca sativa*) on plasma total antioxidant capacity and antioxidant levels in human subjects. *British Journal of Nutrition*, 88 (6): 615-623.

[131] Yang, X. Wei, S. Liu, B. Guo, D. Zheng, B. Feng, L. Liu, Y. Tomás-Barberán, F. A. Luo, L. and Huang, D. (2018). A novel integrated non-targeted metabolomic analysis reveals significant metabolite variations between different lettuce (*Lactuca* sativa. L) varieties. *Horticulture research*, 5 (1): 33.

[132] Alarcón-Flores, M. I. Romero-González, R. Martínez Vidal, J. L. and Garrido Frenich, A. (2016). Multiclass determination of phenolic compounds in different varieties of tomato and lettuce by ultra-high performance liquid chromatography coupled to tandem mass spectrometry. *International journal of food properties*, 19 (3): 494-507.

[133] Pepe, G. Sommella, E. Manfra, M. De Nisco, M. Tenore, G. C. Scopa, A. Sofo, A. Marzocco, S. Adesso, S. and Novellino, T. (2015). Evaluation of anti-inflammatory activity and fast UHPLC–DAD–IT-TOF profiling of polyphenolic compounds extracted from

green lettuce (*Lactuca sativa* L.; var. Maravilla de Verano). *Food Chemistry*, 167: 153-161.

[134] Chiwocha, S. D. Abrams, S. R. Ambrose, S. J. Cutler, A. J. Loewen, M. Ross, A. R. and Kermode, A. R. (2003). A method for profiling classes of plant hormones and their metabolites using liquid chromatography-electrospray ionization tandem mass spectrometry: an analysis of hormone regulation of thermodormancy of lettuce (*Lactuca sativa* L.) seeds. *The Plant Journal*, 35 (3): 405-417.

In: Lactuca: Cultivation and Uses
Editor: Jan Krüger

ISBN: 978-1-53617-729-9


*Chapter 3*

# BIOLOGICAL ACTIVITIES OF *LACTUCA SATIVA* EXTRACTS

***Hayet Edziri*[1], *Fethia Skhiri*[2] *and Maha Mastouri*[1]**

[1]Laboratoire des Maladies Transmissibles et des Substances Biologiquement Actives, Faculté de Pharmacie, Monastir, Tunisia

[2]Laboratory of Genetic Biodiversity and Valorisation of Bioresources, Higher Institute of Biotechnology of Monastir, Tunisia

## ABSTRACT

*Lactuca sativa* var crispa has been more present in the nutrition in the last few years. This vegetable has a high prophylactic effect due to high level of biologically active matters in it. Its strong attractive colors indicate the presence of phenol compounds and therefore good antiradical activity.

Antibacterial, anticoagulant and antioxidant activities of *Lactuca sativa* var. crispa extracts were investigated. The antibacterial activity was evaluated using microwell dilution method against a wide range of clinical isolated bacteria.

Antioxidant activity of *Lactuca sativa* extracts were tested using Ferric-Reducing Antioxidant Power (FRAP) and Catalase Activity Assay

(CAT). Among tested extracts, aerial part extracts showed the best antibacterial activity with Minimal Inhibitory Concentration (MIC) ranged from 0.062 to 1.5mg/ml. In addition, such extracts exhibited the highest anticoagulant and antioxidant activities. The results provided an evidence that the studied vegetable might, indeed, be potential sources of natural antioxidant, anticoagulant and antimicrobial agents.

## INTRODUCTION

Plants play a significant role in maintaining human health and improving the quality of human life. They serve humans well as valuable components of food, such as seasonings and beverages as well as in cosmetics, dyes, and medicines. Many plant extracts prepared from plants have shown to exert biological activity in vitro and in vivo, which justified research on traditional medicinal plants focused on the characterization of their antimicrobial activity (Halliwell and Gutteridge 2007). Large numbers of plants have been screened as a viable source of natural antioxidants including tocopherols, vitamin C, carotenoids and phenolic compounds which are responsible for maintenance of health, to help the human body to reduceoxidative damage and to protect from coronary heart diseases and cancer (Rada and Leto 2008; Conner et al. 2002). Phytochemicals in vegetables can neutralize oxidative agents. Beneficial effects of phytochemicals are believed to be achieved through several mechanisms, such as stimulation of the immune system, modulation of gene expression and hormone metabolism, chelation of transition metals and providing antibacterial and antiviral supports.

The health benefits of vegetables in preventing cancer and cardiovascular diseases are mostly attributed to the quality and quantity of antioxidative components.

*Lactuca sativa var crispa* called romaine lettuce, belongs to the botanical family of Asteraceae. It is a variety of lettuce which grows in a tall head of sturdy leaves with a firm rib down the center. Richly favored romaine lettuce is firm and crisp enough to be heated and served in warm salads. It is an essential ingredient in Tunisia salad and in Mexican and

American caesard salad. Anticonvulsant and sedative-hypnotic effects have been mentioned for the leaves of this plant (Chu et al. 2002). Sayyah et al. (2004) investigated that seeds extract had analgesic and anti-inflammatory activity in rats.

The aim of the present study was to investigate the antibacterial, anticoagulant and antioxidant activities of *Lactuca* sativa *var crispa* leaves extracts growing in Tunisia.

## MATERIALS AND METHODS

### Plant Material

The herb was purchased in June from a local market in Sousse (sahel Tunisia) and the plant aerial parts were authenticated and a voucher specimen was deposited in our laboratory of Faculty of Pharmacy.

### Preparation of Extracts

#### *Methanol Extract*

The leaves herbs were extracted with absolute methanol, in a 1:10 (w/v) ratio of herb to solvent, for 4 h under a continuous reflux set-up in a Soxhlet extractor.

After the extraction, the methanol extracts were clarified by filtering through Whatman # 1 filter paper, followed by centrifugation at 14,000 × g for 5min.

All clarified methanol extracts were stored at -20 prior to experimentation.

***Aqueous Extract***

Samples of leaf of lettuce were extracted with water at 80ºC, in a 1:10 (w/v) ratio, for 4h under continuous shaking. After extraction, the water extracts were clarified by filtering through Whatman # 1 filter paper, followed by centrifugation at 14,000 × g for 5min. All clarified water extracts were stored at -20 C prior to experimentation.

## Total Phenolic Contents

The polyphenol content of the extracts was determined spec-Trophotometrically according to the Folin–Ciocalteu colorimetric method (Zovko et al. 2010), calibrating against catechin standards and expressing the results as mg catechin equivalents per gram of extract (CE)/g extract. Data presented are average of three measurements.

## Total Flavonoids Content

The total of flavonoids contents of *Lactuca* extracts was measured by the colorimetric method of aluminumchlorite. 1mL of extracts were added to a 10mL volumetric flask containing 4mL of H2O and 0.3 mL of 5% sodium nitrite. After 5 min incubation, 0.3mL of sodium chloride (10%)was added to the mixture. After 6min, 2mL of NaOH(1M) were added and the final volume was adjusted to 10 mL with H2O. Finally, the optical density was read by the spectrophotometer at a wavelength of 510nm (Kähkönen et al. 1999).

## Determination of Pigment Content

The procedure was carried out at 4°C and in the dark. Leaf sample (0.25g) were mashed in a pestle and mortar with 80% acetone (v/v).The extract was filtered through two layers of nylon and centrifuged in sealed tubes at 15,000 x g for 5min .The supernatant was collected and the absorbance was read at 663 and 645 nm for chlorophyll a and chlorophyll b, respectively.

The total chlorophyll Chl(a + b) concentration was given in μg $ml^{-1}$ of extract solution according to the equations of Lichtenthaler and Buschmann (2001):

Total Chlorophyll (μg/ml) = 20.2(A645) + 8.02(A663)

## Antibacterial Activity

The MIC (Minimum Inhibitory Concentration) and MBC (Minimum Bactericidal Concentration) were determined on the plant extracts that showed antimicrobial activity, by a broth microdilution method proposed by Novy et al. (2015) with minor modifications. Briefly, 100 μL of Mueller-Hinton Broth (Difco) plus different concentrations of plant extracts were prepared and transferred to each microplate well to obtain serial dilutions of the active extract, ranging from 4 to 512 mg/mL. Then, 10 μL of a fresh culture (final concentration of $1 \times 10^6$ CFU/mL) of test organisms was added. Microplates were incubated at 37°C for 24 h. Wells with no added plant extract were used as a positive growth control. Wells without added bacteria were used as a negative growth control. MIC was defined as the lowest concentration of the extract that restricted the visible growth of microorganism tested.

To determine MBC, 100 $\mu$L from each well that showed no visible growth were re-inoculated on MH agar plates incubated at 37°C for 24 h. MBC and was defined as the lowest extract concentration showing no bacterial and fungi growth.

## Anticoagulant Activity

The anticoagulant activity was evaluated by prothrombin time (PT) and activated partial thromboplastin time (aPTT), as previously described by (Mao et al. 2009). Human Blood were collected from the laboratory of hematology of Fattouma Bourguiba Hospital (Monastir) and then anti-coagulated using 3.8% tri-sodium citrate in a polypropylene container. Then, it was immediately centrifuged for 15min, and plasma was separated, pooled and stored at 4°C until its use. 50µl of extracts was mixed with 100µl of plasma and incubated at 37°C for 5 min at 37°C. Then 200µl of PT assay reagent (rabbit brain extract and calcium chloride)

pre-warmed at 37°C for 15min was added and the clotting time was determined by a coagulometer. For aPTT tests, plasma (100µL) was mixed with 50µL of extracts and 100 of aPTT reagent after that incubated at 37°C for 2min., followed by the addition of 50µl of calcium chloride (0.25mM) and the clotting time was measured by an oagulometer. Normal saline was used as negative control and heparin (1IU/mL) was used as positive control.

## Antioxidant Activity

### *Ferric-Reducing Antioxidant Power Assay*

The antioxidant activity of mushroom extracts was assessed by the FRAP method which allows to measure the reducing power at different concentrations (10, 5, 2.5, 1.25, and 0.62mgmL -1,) mixed with 1mL of a sodium phosphate buffer solution (0.2M, pH 6.6, 2.5mL), 1mL of potassium and 1mL of ferricyanide. After 20 min of incubation at 50°C, 1 mL of 10% trichloroacetic acid (w/v) was added. The mixture was centrifuged and the supernatant was mixed with 2.5mL of ionized water and 0.5mL of freshly prepared solution of ferric chloride (0.1%). The optical density of the reaction mixture was measured using a spectrophotometer at 700nm. High absorption indicates a high reducing power. BHT was used as a positive control (Oyaizu 1986).

***Catalase Activity Assay***

Catalase activity was measured according to Aebi's method (Aebi, 1984). Hydrogen peroxide ($H_2O$) disappearance was monitored kinetically at 240nm for 1min at 25°C. The enzyme activity was calculated using an extinction coefficient of $0.043 Mm^{-1}cm^{-1}$, one unit of activity is equal to 1µmol of $H_2O_2$ destroyed/min/mg protein.

## STATISTICAL ANALYSIS

Data are expressed as mean ± standard deviation (SD). Statistical analysis involved a one way analysis of variance (ANOVA). A value of P less than 0.05 ($p < 0.05$) was considered statistically significant.

## RESULTS AND DISCUSSION

### Total Phenolic Content

Phenolic compounds analysis as can be seen from Table 1, the methanolic extract of *Lactuca* sativa leaves has the best amount of flavonoids (12,24 µg CE/mg) and polyphenols (245,75 µg CE/mg) than the aqueous extract. Furthermore the chlorophyll content in the methanolic extract was higher than the aqueous extract.

**Table 1. Polyphenolic, flavonoids and chlorophyll contents of *Lactuca Sativa* extracts**

| **Extracts** | **Polyphenolic content (µgCE/mg)** | **Flavonoïd content (µgCE/mg)** | **Chlorophyll content ch(a+b) µg/ml** |
|---|---|---|---|
| Aqueous | 100.29 ±3.23 | 4.12 ±0.15 | 45.75±7.15 |
| Methanolic | 245.75±1.54 | 12.24 ±0.23 | 54.87±5.75 |

**Table 2. Antibacterial activity of *Lactuca Sativa* extracts**

| | **MIC(MBC)[a]** | |
|---|---|---|
| Strains | Met | Aq |
| *S. aureus SARM[1]* | 20(50) | 50(100) |
| *Acinetobacter IMP/R* | 20(50) | 100(20) |
| *P. aeruginosa IMP/R* | 20(50) | 100(200) |
| *E. cloacae BLSE* | 20(50) | 50(200) |
| *E. cloacae CI* 13047 | 20(20) | 50(100) |
| *K. pneumoniae ATCC* 13883 | 20(50) | 50(100) |
| *A. baumannii ATCC* 19606 | 20(50) | 50(200) |

Met: Methanolic extract, Aq: Aqueous extract, MIC (Minimum Inhibitory Concentration) and MBC (Minimum Bactericidal Concentration) mg.ml-1.

## Antibacterial Activity

The antibacterial activity of *lactuca staiva var.crispa* leaves extracts are summarized in Table 2. As it is shown, methanol and aqueous extracts exhibited antibacterial activity. But methanolic extract exhibited the best bactericidal activity against the tested strains with minimal bactericidal concentration MIC of 20mg/ml and with MBC of 50mg/ml. This extract exhibits strong antibacterial activity against multidrug resistant bacteria especially. We can attributed this good antibacterial activity to polyphenols in the extracts which was known by its remarkable antimicrobial activity (Lui et al. 2014, Edziri et al. 2019).

In other scientific research we have investigated the antiviral activity of *Lactuca* sativa extract var longiflora (Edziri et al. 2011) and we can conclude that the antibacterial activity results of *Lactuca* staiva var.longiflora extracts was almost the same results than the antibacterial activity of *Lactuca* sativa var.crispa. There aren't any differences between the two varieties.

## Anticoagulant Activities

The anticoagulant activities were measured by PT and APTT. APTT is used to evaluate the coagulation factors such as VIII, IX, XI, XII, and prekallikrein in intrinsic coagulation pathway. PT is used to evaluate the coagulation factors V, VII, and X in extrinsic coagulation pathway (Himou et al. 2017). In our study, the results of PT and aPTT of *lactuca sativa* extracts are summarized in Table 3. The highest prolongations of PT were observed with methanolic extracts with PT of 63.87 s (Table 3) but aqueous extract also showed good anticoagulant activity with PT of 51.9 s, as compared with positive control Heparine (40.8 s) we can conclude that they have almost the same activity. In addition the prolongation of aPTT shows the inhibition of the intrinsic common pathway of coagulation. Furthermore, aqueous and methanolic extracts had moderate anticoagulant activity with aPTT of 45.78 s and 74.43 s, respectively.

**Table 3. Anticoagulant activity of *Lactuca Sativa Var. Crispa* extracts**

| **Compounds Sample** | **Dose** | **aPTT (seconds)** | **PT(s)** |
|---|---|---|---|
| Control | Saline | 34.4 ± 2.87 | 15 ± 2 |
| Heparine | 1 IU/mL | 111.8 ± 4.79 | 40.8 ± 2.48 |
| Met | pure | 74.43± 1.83 | 63.87* ± 4.67 |
| Aq | 1 mg/ml | 45.78± 1.17 | 51.9* ± 2.31 |

N = 3, Values are means ± S.E.M., *p < 0.05 significant difference as compared to control.

There is no literature data on the anticoagulant activity of *Lactuca sativa var.crispa leaves extracts*, so our study is the first one to prove this activity.

We can attribute this activity to the high amount of total polyphenol and flavonoids as demonstrated in others scientific reports (Himou et al. 2017; Félix-Silva et al. 2014).

## Antioxidant Activity

### *Ferric-Reducing Antioxidant Power Assay*

The reducing power of the extracts of the two extracts of *Lactuca* shows an important antioxidant activity with the reducing power assay. The methanolic extract showed the best activity with the lowest $IC_{50}$ (245.32μg.ml-1) (Table 4).

Similar results have also investigated that the reducing power of the common Schizophyllum extract could be due to their electron transfer capacity that could react with free radicals to stabilize and terminate the radical chain (Berker et al. 2007). Apparently, the reducing power of methanolic extracts was usually higher than the other extracts as it was demonstrated in other scientific reports (Kültüret al. 2011; Hadžifejzović et al. 2013).

**Table 4. Antioxidant activity of *Lactuca Sativa Var. Crispa* extracts**

| Extracts | Reducing power $IC_{50}$ μg.ml$^{-1}$ | Catalase (μmol $H_2O_2$ degraded/min/protein) |
|---|---|---|
| Met | 245.32±3.17 | 356.43 ±2.87 |
| Aq | 314.43 ± 2.54 | 210.59± 1.13 |
| BHT | 0.02±0.01 | - |

$IC_{50}$ (mg mL -1): the concentration at which 50% is inhibited, Met: Methanolic extract, Aq: Aqueous extract.

### *Catalase Activity Assay*

In Table 4 we have demonstrated that the activity of the catalase enzyme from the methanolic extract of *Lactuca sativa* leaves (356.43 μmol $H_2O_2$ degraded/min/protein) was higher than the aqueous extracts (210.59 μmol $H_2O_2$ degraded/min/protein).

This decrease in catalase activity or peroxide accumulation hydrogen correlates with cancer metastases (Halliwell and Gutteridge 2007; Tsao et al. 2014). Catalase plays a protective role against cardiovascular diseases. These results confirm the correlation between our results in antioxidant activities and the antioxidant activity of *Lactuca* sativa var crispa extracts by DPPH and ABTS assays wich we have demonstrated in other scientific report (Edziri et al. 2011). This work showed the importance of *Lactuca sativa* in our daily diet.

## Conclusion

On the basis of results in this study, it can be concluded that methanol extract of *Lactuca sativa* had the highest total phenolic and flavonoids and chlorophyll contents than aqueous extract. It showed strong antioxidant, anticoagulant and antibacterial activities than the aqueous extract so these high activities could be important to preserve food products. Considering that anticoagulant activity associated with antioxidant properties could be beneficial for various cardiovascular diseases. However, other research is interesting to determine the active compound in *Lactuca sativa extracts* responsible for the observed biological activities.

## References

Conner, G. E., Salathe, M. and Forteza, R. (2002). "Lactoperoxidase and hydrogen peroxide Metabolism in the airway". *Am J Resp Crit,* 166: 557 - 560.

Changwei, A., Anping, L., Abdelnaser, A.E., Tran, D. X and Shinkichi, T., 2008. Evaluation of antioxidant and antibacterial activities of *Ficus microcarpa* L. fil. Ext. *Food Cont*, 19: 940–948.

Edziri, H., Maha, M., Ammar, S. and Aouni, M. (2008). Antimicrobial, antioxidant, and antiviral activities of Retama raetam (Forssk) Webb flowers growing in Tunisia. *World J Microbiol Biotech,* 24: 2933 - 40.

Edziri, H., Maha, M., Laurent, G., Zine, M. and Aouni, M. (2012). "In vitro evaluation of antimicrobial, antioxidant activities of some Tunisian vegetables". *South Afr Jl of Bot* 78: 252 - 56.

Félix-Silva J., Gomes, J. A. S., and Barbosa, M. Q. (2014). "Systemic and local anti-inflammatory activity of aqueous leaf extract from *Jatropha gossypiifolia* L. (Euphorbiaceae)". *Int J Pharm Phar Sci* 6: 142 - 145.

Hadzifejzovic, N., Kukić-Marković, Jelena S. P., Marina, S. and Adolf, N. (2013). "Bioactivity of the extracts and compounds of *Ruscusaculeatus* L. and *Ruscushypoglossum* L". *Indus Crops Prod*, 49: 407 - 411.

Halliwell, B. and Gutteridge, J. M. C. (2007). *Free Radicals in Biology and Medicine*, 4th. ed. Oxford University Press, Oxford.

Himou, S., 2017. Anticoagulant Activities of Olea Europaea Leaves and Fruit Extract. *Europ Sci J* 13: 1857 - 7881.

Kähkönen, M. P. and Hopia, A. I., (1999). Antioxidant activity of plant extracts containing phenolic compounds. *J Agr Food Chem.* 47: 3954 - 3962.

Kültür, S. (2007). Medicinal plants used in Kirklareli Province (Turkey). *J Ethnopharm* 111: 341 - 364.

Lichtenthaler, H. K. and Buschmann, C. (2001). *Chlorophylls and carotenoids–Measurement and characterization by UV-VIS.* In: Lichtenthaler, H. K. editor. Current Protocols in Food Analyticial Chemistry (CPFA), (Supplement 1). New York, Wiley.

Luis, A., Breitenfeld, L., Ferreira, S., Duarte, A.P. and Domingues, F. (2014) "Antimicrobial, antibiofilm and cytotoxic activities of *Hakea sericea* Schrader extracts". *Phar Mag* 10:213 - 215.

Novy, P., Davidova, H., Serrano R., Pulkrabek, J. and Kokoska, L. (2015). "Composition and Antimicrobial Activity of Euphrasiarostkoviana Hayne Essential Oil". *Evidence-Based Compl Alt Med* 7: 34 - 41.

Rada, B. and Leto, T. L. (2008). "Oxidative innate immune defenses by Nox/Duox family NADPH oxidases". *Contr Microbiol* 15: 164 - 187.

Sayyah, M. and Hadidi, N. (2004). "Analgesic and anti-inflammatory Activity of *Lactuca* sativa seed extract in rats". *J Ethnopharm* 92: 325 - 329.

Tsao, P. S., Heidary, S. and Wang, A. (1998). Protein kinase C-epsilon mediates glucose-induced superoxide production and MCP-1 expression in endothelial cells. *FASEB Journal*, 4: 112 - 115.

Wondrak, G. T. (2009). "Redox-directed cancer therapeutics: molecular mechanisms and opportunities". *Antiox Red Sign,* 11: 3013 - 3069.

Zovko, K. M. and Kremer, D. (2010). "Antioxidant and antimicrobial properties of Moltkia Petraea (Tratt.) Griseb. flower, leaf and stem infusions". *Food Chem Toxicol*, 48: 1537 - 1542.

In: Lactuca: Cultivation and Uses
Editor: Jan Krüger
ISBN: 978-1-53617-729-9


*Chapter 4*

# STRATEGIES FOR GENETIC TRANSFORMATION OF *LACTUCA SATIVA*

***Erum Dilshad[1,*], Hammad Ismail[2], Hasnain Waheed[1], Muhammad Maaz[1] and Zarina Khurshaid[1]***

[1]Department of Bioinformatics and Biosciences,
Faculty of Health and Life Sciences, Capital University of Science and Technology (CUST) Islamabad, Pakistan
[2]Department of Biochemistry and Biotechnology,
University of Gujrat, Gujrat, Pakistan

## ABSTRACT

Lettuce (*Lactuca sativa* L.) is a worldwide vital leafy crop that is cultivated globally. As a result of the speedy progression of human population and their desire to grow, scientists have conducted many studies especially on the genetic transformation or genetic engineering of this plant. The main focus of the current lettuce cultivation programs is to improve horticultural as well as pharmacological characteristics of the plant such as enhancing quality, protection to early bolting, produce disease-free plants and production of important proteins of biopharming.

[*] Corresponding Author's E-mail: dr.erum@cust.edu.pk.

These characteristics can be improved by genetic engineering as well as by adopting different transgenic approaches. The development of efficient and liable tissue culture methods is of prime importance for the successful accomplishment of the genetic transformation of the lettuce. Advances in micropropagation system and transformation methods have aided in increasing the transformation efficacy and constant expression of transgenes in lettuce. Although there is a lot of research that has been done on lettuce with the main purpose to produce pharmaceutically important proteins and vaccines, detailed comprehensive reviews on the basic transformation strategies and major regeneration system of lettuce plant are still missing. Moreover, chloroplast transformation offering unique benefits including high-level expression of foreign protein also needs to be considered. This chapter is focused on the genetic transformation of the lettuce including both nuclear as well as chloroplast transformation system along with major regeneration system and applications in plant biotechnology.

**Keywords**: biopharmaceuticals, chloroplast transformation, genetic transformation, lettuce, tissue culturing

## INTRODUCTION

Genetic transformation is a natural process that involves the transfer of a gene from one organism to another in order to produce variation in biological traits. In order to transcribe transgenic DNA in the new generation, it is essential to transform the germline cell or any other somatic cell of a species. In eukaryotes usually, the bacterium is required as a vector for the transfer of gene of interest. Plant transformation has now become an important method to transform a gene of interest into a particular plant to get the required trait. Stable integration of genes into plant nuclear genome must be done to get the required expression. The basic purpose of the production of transgenic plants are - a) To improve the yield of a crop. b) Improvement of traits. c) Resistance against environmental factors [1].

Plant transformation can be done in two ways, direct or physical and indirect or biological methods. The indirect method is actually the *Agrobacterium*-mediated gene transfer method whereas, direct methods include particle biolistic method or gene gun method of DNA delivery. Whereas electroporation, polyethylene glycol mediated transformation, macro injections, fiber mediated DNA transfer, LASER induced DNA transfer, liposome-mediated DNA transfer and pollen DNA transformation, these all are chemical-based methods of transformation that are actively involved in the transformation of various species [2].

*Agrobacterium*-mediated transformation is a method to manipulate a single gene or many genes to get the desired product and it plays an important *role* in the field of plant biotechnology. *Agrobacterium*-mediated transformation is totally dependent on *Agrobacterium* virulent strains. *Agrobacterium* DNA is used as a vector to transfer the specific gene into the genome of a plant. Today, many horticulturally and agronomically important plants species are transformed by using this bacterium [3]. In some developed countries, many economically important plants species are transformed by using *Agrobacterium*. In the same way, the virulent strains of *Agrobacterium* are responsible for the development of many diseases such as crown gall and hairy root disease in different plant species. Through *Agrobacterium*-mediated transformation, we can enhance the quality and yield of the specific plant species [4].

## *Agrobacterium* Species and Their Range of Hosts

The *Agrobacterium* genus is a wide genus that is divided into several species based on the host range and the type of disease caused. *A. radiobacter* is an avirulent strain that does not have any harmful effect on plant species. *A. tumefaciens* is involved in causing crown gall disease in various plants. *A. rhizogenes* usually causes hairy root disease and *Agrobacterium* species *A. rubi* damages the plant by developing cane gall disease. Few other species have recently been suggested, such as *A. vitis* that causes galls on grapes and other plant species. *Agrobacterium* has the ability to transfer DNA to a wide group of organisms including various monocots and dicots angiosperm and gymnosperm species [5]. *Agrobacterium* can also transform many yeasts, basidiomycetes and ascomycetes. Recently, *Agrobacterium* has been reported to be capable of transferring DNA into a human cell.

## *Agrobacterium tumefaciens*

*Agrobacterium tumefaciens* is a virulent, gram-negative, rod-shaped bacteria that are usually present in the soil. This bacterium damages the plant by causing crown gall disease. *A. tumefaciens* belongs to family Rhizobiaceae that usually contains the nitrogen-fixing symbionts. But unlike nitrogen-fixing symbiont, *A. tumefaciens* is not beneficial for the plant rather it is pathogenic and produces tumor in the plant. *A. tumefaciens* usually attack the wound site of plants. These sites release sugar and phenolic compounds that signal the *A. tumefaciens* bacterium to start the transcription of virulence genes. *A. tumefaciens* contains tumor-inducing (Ti) plasmid and these virulence genes are present on this plasmid [6]. Virulence proteins play an important *rol*e in transcriptional activation of T-DNA, its processing and export while some of the virulence proteins have a specific function in the host.

## Tumour Inducing (Ti) Plasmid

The cell of *Agrobacterium tumefaciens* contains a Ti plasmid (Figure 1). The size of Ti plasmid is about 200-800 kbp. The Ti plasmid contains the genes for inducing tumour in the plant. The transfer DNA (T-DNA) is present in the T-region of Ti plasmid. The size of T- region on Ti plasmid is approximately 10-30kbp. Thus, the T-region is less than 10% of the Ti plasmid. Some Ti plasmids contain only one T- region while some contain multiple T-regions. Usually, *Agrobacterium* transfers its T-DNA to the plant in order to induce the production of food molecules for bacterium

[7]. *Agrobacterium tumefaciens* attack on wound plant. The wound site releases phenolic compounds that are recognized by the receptor (*virA*) that is present on the bacterium. This *virA* detects the chemical signals from the wounded plant. The binding of phenolic compounds activates the *virA* which can activate another protein *virG* through phosphorylation. This *virG* protein is inactive without phosphorylation. This *virG* activates the transcriptional process of *vir* genes present on the Ti plasmid. The transcription of *vir* genes produces two other proteins known as *virD1* and *virD2*. These *virD1* and *virD2* proteins have endonuclease ability and cleave the 25 bp sequence repeats at the end of T-DNA. As a result, the single strand of T-DNA molecule is produced which attaches the *virD2* at one end. This *virD2* directs the T-DNA to be transferred to the plant cell in a process similar to bacterial conjugation. Inside a plant cell nucleus, the T-DNA has the ability to integrate into the genome of the plant. This T-DNA contains the number of genes that are being expressed in the plant cell. Some gene products are responsible for the production of plant growth hormone-like auxin and cytokinin. These hormones trigger the plant cell to divide more rapidly. With the action of these hormones, the division of plant cell rapidly increases that ultimately leads to tumor formation (crown gall tumor). The T-DNA also contains some genes that are responsible for the production of unusual amino acid derivatives known as opine. These opines are a good source of carbon and nitrogen and hence utilized by the *Agrobacterium* for its growth. But in *Agrobacterium*-mediated transformation, the T- DNA is cleaved from the Ti plasmid and this T-DNA incorporate into another plasmid to form a recombinant [8]. This repetitive sequence present on T-DNA is replaced with the gene of interest. This T-DNA do not contain any tumor-inducing genes. Now, this recombinant containing the gene of interest is introduced into a suitable plant to get the desired product. The introduction of T-DNA into the specific site of the plant genome is very important. The T-DNA must be incorporated within the transcriptionally active regions that contain promoter and enhancer. This specific region is involved in the transcription of the gene of interest. Nowadays, many economically and industrially

important plants are transformed by *Agrobacterium*-mediated transformation.

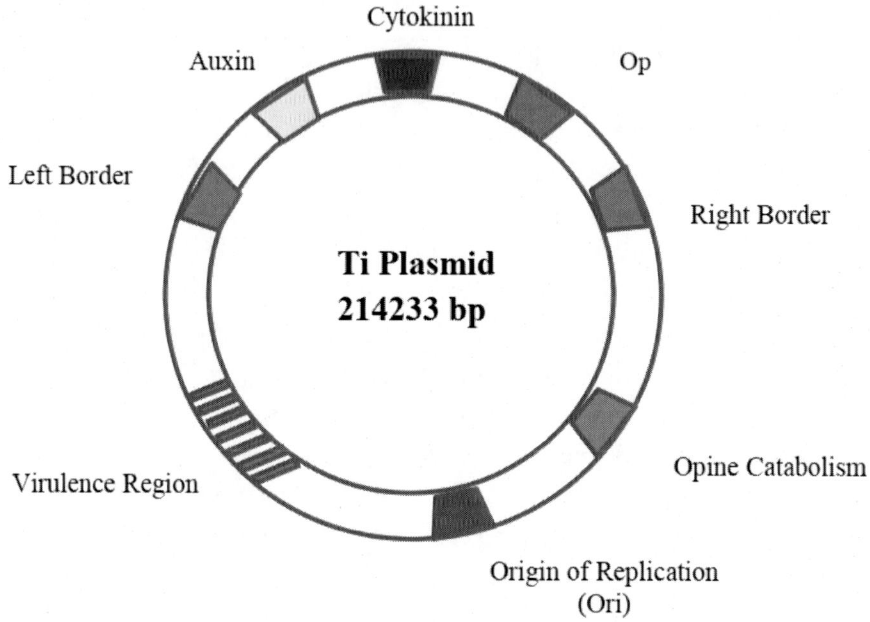


Figure 1. Ti plasmid of *Agrobacterium tumefaciens.*

## *LACTUCA SATIVA*

*Lactuca sativa* is a well-known edible crop that belongs to the *Lactuca* genus and the family Asteraceae. Tshilai, lettuce and salad are the public names used for *L. sativa*. Lettuce is a leafy crop that is native to western Asia and the eastern Mediterranean region. Ancient Romans and Greeks were well familiar to the therapeutic, medicinal and nutritional value of the lettuce. Lettuce has been growing in China since the 5$^{th}$ century. There were a lot of varieties of lettuce introduced in America by Cristopher Columbus in 1493 [9].

Lettuce is an annual or biennial lactiferous herb. It ranges in height from 15-30 cm, with colorful leaves from red and yellow to dark green. There is a lot of varieties of lettuce depending upon the texture and shape [10]. It is found rich in vitamin A, K and also a rich source of potassium to the human population. It is also low in calories, good tasted with high nutritional value. There are a lot of lettuce varieties that have recently been investigated and found to contain phenolic compounds along with anti-oxidants [11]. The nutritional benefits of the lettuce are the presence of phenolic compounds, vitamin C and a good amount of fibre content [12]. In folk medicine, the seeds of lettuce were used for the treatment of cough, asthma and utilized as an analgesic as well. Iceberg lettuce is very commonly used (especially in restaurants). The number of antioxidants phenolic compounds in lettuce may vary within varieties due to processing, growing practices and storage conditions. Therefore, different transformational strategies are applied to enhance the quality, quantity and shelf life of lettuce [13]. In this chapter, we discussed the *Agrobacterium*-mediated genetic transformation of the nuclear genome as well as some strategies of chloroplast genetic transformation of *Lactuca sativa.*

## AGROBACTERIUM-MEDIATED TRANSFORMATION IN *LACTUCA SATIVA*

Nowadays, many transformations have been done in the *Lactuca sativa* genome in order to get beneficial effects [14]. Different transformations are done in lettuce to get the disease-free plant, increasing nutritional value, and get resistant to many pests. Some of the details are given in Table 1.

**Table 1. Agrobacterium tumefaciens mediated genetic transformation of *Lactuca sativa L.***

| Genotype | Detection | Plasmid vector | Function | Reference |
|---|---|---|---|---|
| Grand Rapids | PCR, RT-PCR | pPCV002-ABC | Increased flavonoids | [15] |
| Grand Rapids | PCR, RT-PCR | pPCV002-CaMVC | Increased flavonoids | [16] |
| *Lactuca sativa* | PCR-Southern, Western blot analyses | pBinU-Omega-HisNP1 | Broad-spectrum disease resistance | [17] |
| *Lactuca sativa* cv. Solan Kriti) | | pCambia | Fungal resistance | [18] |
| Grand rapids | PCR, PCR-Southern, RT-PCR | pCAMBIA1301 | FMD | [19] |
| VitoriadeVerao | PCR | PG35SHBsAg | HBV | [20] |
| Glass lettuce | PCR, RT-PCR | pBI121-NA | AI | [21] |
| *Lactuca sativa* | PCR, Northern blot, Immunoblot, $GM_1$-ELISA Southern blot, western blot, $GM_1$ BA | pMYV514 | Cholera toxin | [22] |
| Snezhinka et al. | PCR, RT-PCR | pCB063, pCB064 | Tuberculosis | [23] |
| Green Wave | PCR, Southern blot, RT-PCR, ELISA, | pBIF-V | Plague | [24] |
| *Lactuca sativa* | PCR, Northern blot, Western blot, Laser, Scanning lonfocal Microscopy | pCV1, pCV2, pCV12 | SARS | [25] |
| *Lactuca sativa* | Laser scanning lonfocal Microscopy pcr northern blot,immune blot ELISA GM1-ELISA | PMYO51 | HEAT-labile Enterotoxin | [26] |
| *Lactuca sativa* | Western blot, FAS, Mouse feeding assay | pBI121-*espA* | Enterohaemorragc *E. coil* | [27] |
| Zhouye | PCR, Southern blot, RT-PCR | p35S-2300-twinT-DNA::pil-msCT::noster | sCT | [28] |
| Grand rapids | PCR, ELISA, ifferential Spectroscopy | pGBIVHbTα1hLf, pGBIRVHbTα1hLf | Human lactoferrin and Thymosin | [29] |
| *Lactuca sativa* | PCR, RT-PCR | pBI-121-NK | nattokinase | [30] |
| *Lactuca sativa* | GUS assay, RT-PCR, ELISA | pSFIFN-α | ChIFN-α | [31] |
| Italian Yearly Late Bolt | PCR, RT-PCR, HPLC-ELSD | p2301-GMP-myc | Vitamin C | [32] |

| Genotype | Detection | Plasmid vector | Function | Reference |
|---|---|---|---|---|
| *Longifolia* L am. | PCR, RT-PCR, HPLC | pCAMBIA2300-35S::LsHPT::NOS | Vitamin E | [33] |
| Grand rapids | PCR, Southern blot, RT-PCR, AA | pBI121-lrp | Lysine | [34] |
| Romaine | RT-PCR, HPLC, *Lactobacillus* case fermentation | pFSndt5 100-Atpsy-fole | Carotenoids and Folic Acid | [35] |
| Evola | PCR, RT-PCR | pBIPTA | Resistance to Aphids | [36] |
| Kaiser | DAS-ELISA, Western blot, Southern blot, Northern blot | pYK23 | Resistance to Mirafiori virus | [37] |
| Verônica | PCR, RT-PCR | pCambia xDc | Resistance to Sclerotiniasclerotiorm | [38] |
| Chongchima | Histochemical GUS staining, Southern blot, Northern blot, Drought and cold stress tests | pCUMB | Resistance to drought and cold | [39] |

## Transformation with Herbicide Resistance Genes

The alteration of various characters of lettuce that are agronomically important is usually done with the recombinant DNA transformation technology. *A. tumefaciens* facilitated the transformation of bialaphos resistance gene *(bar)* into the cultivar Evola of lettuce [40]. T1 and T2 both plant generations showed stable expression of the resistance gene. The plants that were cultivated in medium containing 5mg/L glufosinate ammonium were found resistant when sprayed with 300mg/L of the herbicides. In another study, *Agrobacterium tumefaciens* containing a gene for the enzyme 5-enolpyruvyl shikimate-3-phosphate synthase (*EPSPS*) rendering tolerance against broad-spectrum herbicide glyphosate was used for transformation. As a result of this, 6 glyphosate resistant transformed plants of lettuce were produced [41].

Curtis et al. (1999) transformed a specific gene nitrate reductase (*nia2*) into the lettuce cultivars Flora, Luxor, Cortina and Evola. Nitrate reductase enzymatic assay and Southern hybridization confirmed the transgenic status of plants. There is no decline in the nitrate content in the transgenic plant as compared to field-grown plants [42]. Dubois et al. (2005) have done the transformation under the cont*rol* of 35S promoter of cauliflower mosaic virus of the *Lactuca sativa* cv. Jessy has the same nitrate reductase gene *nia2*. They suggested that the expression can be off if the nitrate reductase mRNA gene is silent [43].

## Transformation for Disease Resistance

Valimareanu and his colleagues reported the genetic transformation in *Lactuca sativa*. In his experiment, he tried to make a disease-free lettuce plant. Because there are many insects that damage the lettuce crop every year. He successfully introduced sap-sucking insect-resistant *Agrobacterium tumefaciens* LBA 4404 (1065) strain containing hypervirulent pTOK47 plasmid and a binary vector pMOG23. Salmon *ct* (calcitonin), *pta (Pinellia ternata* agglutinin*)* and *cgrp* (calcitonin gene-related protein) genes were integrated into *Lactuca sativa* plant. These genes usually involved in the synthesis of calcitonin. *Pinellia ternata* agglutinin is a gene that is also responsible for the synthesis of lectin. Lectin is a chemical compound that has a lethal effect on homopteran and orthopteran insects. Lectin acts as a defence system for various plant bodies. So inducing these two chemical genes into the lettuce genome will get insect-resistant *Lactuca sativa* plant [44].

## Transformation to Enhance the Nutritional Value

*Lactuca sativa* is a rich source of vitamin C so there are many transformations done in lettuce plant to enhance its nutritional value. WeiPeng and co-workers in 2011, tried to improve vitamin C content in *Lactuca sativa*. The GDP-mannose pyrophosphorylase (GMP) gene is responsible for the enhancement of vitamin C content. They cloned the (GMP) gene with specific primers from the cDNA of Arabidopsis thaliana. CaMV 35S promoter, MYC sequence and NOS terminator were connected to GMP gene in order to develop a GMP-containing expression cassette, after that, it was transferred into pCAMBIA2301 expression vector then p2301-GMP-myc. *L. sativa* was transformed with Agrobacterium. By using the techniques of PCR and RT-PCR it was confirmed that the GMP was efficiently integrated and expressed within *L. sativa*. Now, the HPLC-ELSD technique was used to measure the content of vitamin C in transgenic *L. sativa*. The results indicated that the transgenic *L. sativa* has a high content of vitamin C as compared to its wild type [45]. So this was cleared that the overexpression of GMP gene will increase the vitamin C content in *L. sativa*.

Nowadays, research has been done on lettuce to improve its shelf life. For this purpose, Sun et al. (2006) introduced a step to increase the shelf life of lettuce, they successfully cloned the miraculin gene that is isolated from the Richadella dulcifica (a West African shrub) and introduced this gene into the Lectuca sativa with the help of A. tumefaciens [46]. Cho et al. (2005) improved the tocophe*rol* composition in the lettuce cv. Chongchima. A gene coding γ-tocophe*rol* methyltransferase from Arabidopsis thaliana was used to increase enzymatic potential and alteration of γ-tocophe*rol* to the more powerful form. In developing countries, the improvement of plant contents is a big step toward the improvement of human nutrition [47].

Micro and macro-elements play a significant *rol*e in oral function, enzyme activity and in general health of an individual. Xiaofeng et al. (2002) investigated the accumulation of increased zinc content in the cv. Salinas 88. The βcDNA is a mouse metallothionein mutant DNA that was inserted through A. tumefaciens mediated transformation. The amount of zinc in transformed plants was up to 0.4mg/g dry weight, significantly higher than in field-grown plants [48]. Goto et al. (2000) successfully transformed the cv. Green leaf through a unique plasmid containing kanamycin-resistant gene (nptII) and CaMV 35S promoter and soybean ferritin cDNA. The plant gained a weight of about 27-42% that has been observed in transgenic plants during their early stages of development and the rate of photosynthesis was also increased. It was observed that the iron content was 1.2-1.7 times greater in transformed plants as compared to the original untransformed plant [49].

## Transformation to Enhance Secondary Metabolites

*Lactuca sativa* is an edible crop along with some medicinal importance. There are many secondary metabolites like flavonoids that are produced by *L. sativa*. Ismail and his coworkers tried to enhance the anti-oxidant potential, analgesic, anti-inflammatory and anti-depressant activity of Lectuca sativa. For this purpose, they transformed *L. sativa* with *rol* C gene. This *rol* C gene is responsible for the expression of anti-oxidant, anti-inflammatory, analgesic and anti-depressant effect. Agrobacterium tumefacien strain along with plasmid was used for the purpose of transformation. The *rol* C gene was present in the coding region of T-DNA of plasmid. Explant of *L. sativa* contained the node and internode region for transformation. After infection with Agrobacterium tumefacien the plasmid containing the *rol* C gene became integrate into the *L. sativa* genome. For the confirmation of plasmid integration, PCR technique was used. The *rol* C gene completely expressed in the transformed plants [50] as compared to untransformed plants, it was cleared that the transformed *L. sativa* had an increase in anti-oxidant production as well as enhanced analgesic, anti-inflammatory and anti-depressant activities.

Niki et al. (2001) introduced the gene responsible for the synthesis of pumpkin gibberellin 20 oxidase into the lettuce cv. Vanguard. Southern blot analysis confirmed single gene copy and that transgene was dominant and stable and usually segregated through a Mendelian pattern. Plants exhibited reduced concentrations of GAl and GA4 and dwarf morphology but increased concentrations of GA17 and GA25 [51]. Park et al. (2005) introduced a gene known as late embryogenesis abundant gene (lea) into lettuce by using the technique of Agrobacterium-mediated transformation for drought and saline tolerance [52].

## Production of Important Biopharmaceuticals

As new medical applications are being developed, the demand for the production of pro-insulin is increasing worldwide. Pro-insulin is a prohormone that is the precursor of insulin production in the beta cells of the islets of Langerhans, a specific region of the pancreas. *Lectuca sativa* has the ability to supply this rising demand as a green bioreactor. Mohebodoni and his coworkers worked on the production of pro-insulin form *Lactuca sativa* by using a method of *Agrobacterium*-mediated transformation. They construct a DNA that contained immunoglobin G-binding protein. A gene of *Staphylococcus aureus* (*ProA*) fused to a sequence encoded pro-insulin (*Pins*) was transformed in the *Lactuca sativa*. For the purpose of transformation, *Agrobacterium tumefaciens* was used. In order to verify the integration of *ProA-Pins* gene into the genome of *L. sativa*, the PCR method along with Southern blotting was used. The expression of *ProA* pins fusion gene in *L. sativa* was confirmed by using the technique of reverse transcription PCR. The *ProA* fusion protein was obtained from the leaves of the lettuce plant. It was confirmed that a high percentage of *ProA* fusion protein was produced from the transformed *L. sativa* [53]. This work demonstrated that Pro-insulin produced by plant-based expression system is a low cost and effective approach for the treatment of diabetes.

Negrouk et al. (2005) successfully introduced an important pharmaceutical antibody IgG1 by transforming the lettuce plant. For this purpose, they took commercially available lettuce, the head of the lettuce were infiltrated with *A. tumefaciens* and allowed to incubate at 25°C with a photoperiod of 16 h for 34 days. This type of protocol produced 20-80 mg of functional antibody per kg of lettuce leaves [54]. Joh et al. (2005) did the same work by using the expression of gene β-glucuronidase as a biomarker for the transformation. Lettuce leaf discs of the cultivars, the two plants green forest and heart delight was treated with *A. tumefaciens* in a vacuum and incubated for about 72 hours at 22°C and placed in darkness. About 0.16% of *gus* protein was produced based on the dry weight of tissue. It is thought that the leaf disk incubated in continuous light will produce rapid protein synthesis but the final protein product is similar to the samples that were incubated in dark [55].

## CHLOROPLAST TRANSFORMATION

The chloroplast and mitochondria have their own genome and their genome is much simpler as compared to nuclear genome present in the nucleus of the cell. It is believed that both organelles are evolved from prokaryotes during evolution. It is also evident that many proteins that perform vital functions in both these organelles are encoded by the nuclear genes and then transported to the organelle [45].

Most of the higher plants have about approximately 100 chloroplasts per leaf cell. Each chloroplast contains 100 copies of chloroplast DNA genome. The genome of the chloroplast is double-stranded circular DNA that is present in the stroma. Majority of chloroplast genomes are in size of 120-160 kbp and contain about 120-140 genes. About 100 chloroplast genes are known to code for proteins [50].

### Vector Design for Chloroplast Transformation

The vector for chloroplast transformation is based on selectable marker gene *aadA* that provides resistance to antibiotic spectinomycin. The single foreign gene is fused to regulatory sequences which in turn are flanked on either side by chloroplast DNA [56]. In the construct for expression of multiple genes, the selectable marker is bataine-aldehyde dehydrogenase (*badh*) gene. It is flanked by a promoter and the multiple transgenes are flanked by a terminator. At both ends, chloroplast DNA sequences are present. In between the transgenes, there are ribosomes binding sites to ensure efficient translation [57].

## Introduction of a Foreign Gene into the Chloroplast Genome

Most of the methods are used for introducing the foreign gene into the nuclear genome but these methods are not useful for chloroplast transformation [58]. The most successful and efficient method for inserting foreign genes into chloroplasts is particle gun bombardment. After the bombardment, homologous recombination occurs between the chloroplast DNA sequences on the vector and those of on the genome [59]. This is site-specific integration and thus avoids the frequent problem associated with the random insertion of foreign genes into the nuclear genome. The regenerated plants derived from modified plastome (chloroplast genome) are regarded as the transplastomic plants [60].

## Chloroplast Transformation of *Lactuca sativa* L.

There are a lot of advantages of the transformation of the plastid genome over nuclear genome transformation [61]. It is estimated that about 10,000 copies of the plastid genome are present per cell [62]. This is a very high level of ploidy that ultimately results in the high level of the expression of transgene so that foreign proteins are about 40% of the total soluble cellular proteins [63]. Due to the high expression of the transgene in the chloroplast, this transformation is very beneficial for metabolic engineering, for the use of plants to get biopharmaceuticals and for resistance management [64]. For these applications, this transformation method will play an important *rol*e in the near future. For the production of pharmaceutically important compounds from tobacco, the main hurdle is the high content of nicotine and toxic alkaloid compounds. The protein production from the majority of the plants is usually from the non-green tissues of the plant body such as root (carrot), micro-tuber (potato) and fruit (tomato). The GFP protein accumulated in potato tuber amyloplast is 100 fold less than the leaf chloroplast. Betaine aldehyde dehydrogenase in chromoplast of the carrot, expressed from a transgene of plastid that is about 74.8% of the level observed in leaf chloroplast [65].

These findings suggested that the tissues of plant that contain chloroplast would be very effective for the production of the protein of interest from plastid transgenes and also play a vital *role* in the production of pharmaceutical compounds. However, the technique in which the chloroplast is transformed is not yet available for edible crops. However, hepatitis B virus subunit has been produced by doing nuclear transformation in *Lectuca sativa* but now this product is under clinical trials [66].

After plantation, lettuce has the ability to grow very quickly and can be harvested within a few weeks. The gene that is integrated into plastid move to the nucleus is also reported in various researches. But there is a low frequency of pollen that are derived from transplastomic plants that carry the transgene and integrate into the genome of plastid [64]. Furthermore, lettuce is best for indoor cultivation by the hydro-culture system. A group of scientists developed the plastid transformation system for lettuce, they transformed DNA that contains a spectinomycin-resistant gene (*aadA*) and the elements that are responsible for the regulation of expression, that is flanked by two adjacent lettuce plastid genome sequences and insert in between the *rbcL* and *accD* genes. According to the report, an average of about 1 transplastomic lettuce plant per bombardment was produced. It was shown that the transgene-encoded GFP was expressed by the chloroplast of lettuce and it was about 36% of the total soluble protein. The $T_0$ plants were fertile and the progeny produced had a stable transgene in the genome of the chloroplast. This system is now best for the production of pharmaceuticals, different vaccines and various antibodies in plants [64].

Furthermore, future development in edible vaccine production from lettuce requires exploring a selectable marker safe for consumption by humans or deletion of bacterial *aadA* gene being used here. This grouping of plastid transformation with expertise letting exclusion of the marker gene or antibiotic-free selection [67] would improve the development of plant molecular breeding.

## CONCLUSION

Lettuce (*Lactuca sativa L.*) is a chief garden-fresh green plant and goes to the *Asteraceae* family (Compositae). The emphasis of contemporary lettuce breeding purposes is to increase agricultural features such as quality and to develop resistance against diseases and pests. These features may be enhanced by means of biotechnology and gene transfer approaches.

Industrial and agricultural applications have taken priority in research conducted in the past decade. Improvements in genetic engineering technology have helped to improve transformation proficiency and stable transgene expression in lettuce. So far, the transgenic research conducted has mainly focused on the production of important biopharmaceuticals using lettuce as a bioreactor and also to improve its nutritional quality and physiological value and agronomic traits. The technology of plastid transformation and its applications are likely to offer novel potentials for the usage of eatable green crops as an industrial unit for biopharmaceuticals.

## REFERENCES


[1] Council, N. R. (1987). Gene transfer methods applicable to agricultural organisms, In *Agricultural Biotechnology: Strategies for National Competitiveness*, pp 156-175, National Academies Press (US).

[2] Bowser, A. and Shanley, L. (2013). New visions in citizen science. *Case Study Ser*, 3: 1-53.

[3] Jensen, P. E. and Leister, D. (2014). Chloroplast evolution, structure and functions. *F1000Prime Reports*, 6 (40): 1-14.

[4] Keck, R. W. Dilley, R. A. Allen, C. F. and Biggs, S. (1970). Chloroplast composition and structure differences in a soybean mutant. *Plant Physiology*, 46 (5): 692-698.

[5] Borner, T. (2017). The discovery of plastid-to-nucleus retrograde signaling-a personal perspective. *Protoplasma*, 254 (5): 1845-1855.

[6] Staehelin, L. A. (2003). Chloroplast structure: from chlorophyll granules to supra-molecular architecture of thylakoid membranes. *Photosynthesis Research*, 76 (1-3): 185-196.

[7] Schünemann, D. (2004). Structure and function of the chloroplast signal recognition particle. *Current Genetics*, 44 (6): 295-304.

[8] Goodenough, U. W. and Levine, R. (1970). Chloroplast structure and function in ac-20, a mutant strain of Chlamydomonas reinhardi: III. Chloroplast Ribosomes and Membrane Organization. *The Journal of Cell Biology*, 44 (3): 547-562.

[9] Anilakumar, K. Harsha, S. and Sharma, R. (2017). Lettuce: a promising leafy vegetable with functional properties. *Defence Life Science Journal*, 2 (2): 178-185.

[10] Ismail, H. (2016). Genetic transformation of *Lactuca sativa* L. for the production of pharmaceuticals, In *Biochmistry*, pp 26-51, Quaid-i-Azam University Islamabad, Pakistan, HEC.

[11] Llorach, R., Martínez-Sánchez, A., Tomás-Barberán, F. A., Gil, M. I. and Ferreres, F. (2008). Characterisation of polyphenols and antioxidant properties of five lettuce varieties and esca*role*. *Food Chemistry*, 108 (3): 1028-1038.

[12] Nicolle, C., Carnat, A., Fraisse, D., Lamaison, J. L., Rock, E., Michel, H., Amouroux, P. and Remesy, C., (2004). Characterisation and variation of antioxidant micronutrients in lettuce. *Journal of the Science of Food and Agriculture*, 84 (15): 2061-2069.

[13] Mulabagal, V. Ngouajio, M. Nair, A. Zhang, Y. Gottumukkala, A. L. and Nair, M. G. (2010). In vitro evaluation of red and green lettuce (*Lactuca sativa*) for functional food properties. *Food Chemistry*, 118 (2): 300-306.

[14] Wang, S. Uddin, M. I. Tanaka, K. Yin, L. Shi, Z. Qi, Y. Mano, J. i. Matsui, K. Shimomura, N. and Sakaki, T. (2014). Maintenance of chloroplast structure and function by overexpression of the rice monogalactosyldiacylglyce*rol* synthase gene leads to enhanced salt tolerance in tobacco. *Plant Physiology*, 165 (3): 1144-1155.

[15] Ismail, H. Dilshad, E. Waheed, M. T. and Mirza, B. (2017). Transformation of lettuce with *rol* ABC genes: extracts show enhanced antioxidant, analgesic, anti-inflammatory, antidepressant, and anticoagulant activities in rats. *Applied Biochemistry and Biotechnology*, 181 (3): 1179-1198.

[16] Ismail, H. Dilshad, E. Waheed, M. T. Sajid, M. Kayani, W. K. and Mirza, B. (2016). Transformation of *Lactuca sativa* L. with *rol* C gene results in increased antioxidant potential and enhanced analgesic, anti-inflammatory and antidepressant activities *in vivo*. *3 Biotech*, 6 (2): 215.

[17] Song, D. Xiong, X. Tu, W. Yao, W. Liang, H. Chen, F. and He, Z. (2017). Transfer and expression of the rabbit defensin NP-1 gene in lettuce (*Lactuca sativa*). *Genetics and Molecular Research*, 16 (1): 1-9.

[18] Sharma, S. and Srivastava, D. (2017). Agrobacterium-mediated fungal resistance gene transfer studies pertaining to antibiotic sensitivity on cultured tissues of lettuce (*Lactuca sativa* L. cv. Solan kriti). *International Journal of Current Microbiology and Applied Sciences*, 6: 1687-1698.

[19] Deng, X. Zhou, Y. and Chang, J. (2007). Establishment of genetic transformation system and transgenic studies in lettuce (*Lactuca sativa* var. capatata). *Acta Botanica Yunnanica*, 1 (15): 1-10.

[20] Marcondes, J. and Hansen, E. (2008). Transgenic lettuce seedlings carrying hepatitis B virus antigen HBsAg. *Brazilian Journal of Infectious Diseases*, 12 (6): 469-471.

[21] Fan, Y. Li, J. Ruan, Y. and Liu, C. (2012). Study on introduction of avian influenza antigen gene NA into lettuce. *Hunan Agricultural Science*, 5: 13-16.

[22] Jing, J. Ji, J. Wang, G. Wang, P. and Zhang, C. (2007). Establishment of genetic transformation system of lettuce by *Agrobacterium tumefaciens* and introduction of HIV gap-gp120 gene. *Tianjin Agricultural Sciences*, 13 (1): 14-17.

[23] Huy, N. X. Yang, M. S. and Kim, T. G. (2011). Expression of a cholera toxin B subunit-neutralizing epitope of the porcine epidemic

diarrhea virus fusion gene in transgenic lettuce (*Lactuca sativa* L.). *Molecular Biotechnology*, 48 (3): 201-209.

[24] Rosales-Mendoza, S. Soria-Guerra, R. E. Moreno-Fierros, L. Alpuche-Solis, Á. G. Martínez-Gonzalez, L. and Korban, S. S. (2010). Expression of an immunogenic F1-V fusion protein in lettuce as a plant-based vaccine against plague. *Planta*, 232 (2): 409-416.

[25] Li, H.-Y. Ramalingam, S. and Chye, M.-L. (2006). Accumulation of recombinant SARS-CoV spike protein in plant cytosol and chloroplasts indicate potential for development of plant-derived oral vaccines. *Experimental Biology and Medicine*, 231 (8): 1346-1352.

[26] Kim, T. G. Kim, M. Y. Kim, B. G. Kang, T. J. Kim, Y. S. Jang, Y. S. Arntzen, C. J. and Yang, M. S. (2007). Synthesis and assembly of *Escherichia coli* heat-labile enterotoxin B subunit in transgenic lettuce (*Lactuca sativa*). *Protein Expression and Purification*, 51 (1): 22-27.

[27] Luan, J. Chen, M. Li, Z. and Wang, P. (2009). EspA expressed in lettuce stimulated high immunisation and had a protective effect on HeLa cells. *Food and Agricultural Immunology*, 20 (3): 207-219.

[28] Cui, L. Lu, L. Ye, J. Pang, Y. Shen, G. Zhao, L. and Tang, K. (2009). Development of functional lettuce for preventing and curing osteoporosis. *Journal of Shanghai Jiaotong University-Agricultural Science*, 27 (5): 465-474.

[29] Meng, Z. Zhang, R. Liu, D. Yan, X. Geng, Q. and Guo, S. (2005). Expression of human lactoferrin, thymosin fusion gene in lettuce. National symposium abstract. *Crop Biotechnology Mutagenesis Technology*, 25: 1.

[30] Tian, X. (2007). *A study of transformation of nattokinase into lettuce.* MD Thesis Lanzhou University, 16: 10-20.

[31] Li, S. De-gang, Z. Yong-jun, W. and Yi, L. (2008). Transient expression of chicken alpha interferon gene in lettuce. *Journal of Zhejiang University Science B*, 9 (5): 351.

[32] Wang, W. Guo, X. and Tang, K. (2011). Transformation of GDP-mannose pyrophosphorylase gene from *Arabidopsis thaliana* L. into

*Lactuca sativa* L. *Journal of Shanghai Jiaotong University-Agricultural Science*, 29 (2): 43-49.

[33] Ren, W. Zhao, L. Wang, Y. Cui, L. Tang, Y. Sun, X. and Tang, K. (2011). Overexpression of homogentisate phytyltransferase in lettuce results in increased content of vitamin E. *African Journal of Biotechnology*, 10 (64): 14046-14051.

[34] Li, X. (2006). Expression and Inheritance of Lysine-Rich Protein Gene in Lettuce (*Lactuca sativa* L.). *Chinese Journal of Applied and Environmental Biology*, 12 (4): 472.

[35] Fu, X. Guo, X. You, L. Tang, Y. Cheng, H. Wang, G. and Tang, K. (2012). Co-transformation of Atpsy and folE genes elevated carotenoids and folic acid contents in romanie lettuce. *Journal of Shanghai Jiaotong University-Agricultural Science*, 30 (6): 22-35.

[36] Ahmed, M. Akhter, M. Hossain, M. Islam, R. Choudhury, T. Hannan, M. Razvy, M. and Ahmad, I. (2007). An efficient Agrobacterium-mediated genetic transformation method of lettuce (*Lactuca sativa* L.) with an aphidicidal gene, Pta (*Pinellia ternata* Agglutinin). *Middle-East Journal of Scientific Research*, 2 (2): 155-160.

[37] Kawazu, Y. Fujiyama, R. and Noguchi, Y. (2009). Transgenic resistance to Mirafiori lettuce virus in lettuce carrying inverted repeats of the viral coat protein gene. *Transgenic Research*, 18 (1): 113-120.

[38] Dias, B. Cunha, W. Morais, L. Vianna, G. Rech, E. De Capdeville, G. and Aragao, F. (2006). Expression of an oxalate decarboxylase gene from Flammulina sp. in transgenic lettuce (*Lactuca sativa*) plants and resistance to *Sclerotinia sclerotiorum*. *Plant Pathology*, 55 (2): 187-193.

[39] Vanjildorj, E. Bae, T. W. Riu, K. Z. Kim, S. Y. and Lee, H. Y. (2005). Overexpression of ArabidopsisABF3 gene enhances tolerance to droughtand cold in transgenic lettuce (*Lactuca sativa*). *Plant Cell, Tissue and Organ Culture*, 83 (1): 41-50.

[40] Mohapatra, U., McCabe, M. S., Power, J. B., Schepers, F., Van Der Arend, A. and Davey, M. R. (1999). Expression of the bar gene

confers herbicide resistance in transgenic lettuce. *Transgenic Research*, 8 (1): 33-44.

[41] Nagata, R. T. Dusky, J. A. Ferl, R. J. Torres, A. C. and Cantliffe, D. J. (2000). Evaluation of glyphosate resistance in transgenic lettuce. *Journal of the American Society for Horticultural Science*, 125 (6): 669-672.

[42] Curtis, I. Power, J. Blackhall, N. De Laat, A. and Davey, M. (1994). Genotype-independent transformation of lettuce using *Agrobacterium tumefaciens*. *Journal of Experimental Botany*, 45 (10): 1441-1449.

[43] Dubois, V. Botton, E. Meyer, C. Rieu, A. Bedu, M. Maisonneuve, B. and Mazier, M. (2005). Systematic silencing of a tobacco nitrate reductase transgene in lettuce (*Lactuca sativa* L.). *Journal of Experimental Botany*, 56 (419): 2379-2388.

[44] Zoschke, R. and Bock, R. (2018). Chloroplast translation: structural and functional organization, operational cont*rol*, and regulation. *The Plant Cell*, 30 (4): 745-770.

[45] Turkec, A. (2000). New transformation system: Chloroplast transformation. *Anadolu*, 10 (1): 8796.

[46] Sun, H. J. Cui, M. l. Ma, B. and Ezura, H. (2006). Functional expression of the taste-modifying protein, miraculin, in transgenic lettuce. *FEBS Letters*, 580 (2): 620-626.

[47] Cho, E. A. Lee, C. A. Kim, Y. S. Baek, S. H. Reyes, B. G. and Yun, S. J. (2005). Expression of γ-tocophe*rol* methyltransferase transgene improves tocophe*rol* composition in lettuce (*Latuca sativa* L.). *Molecules and Cells*, 19 (1): 1.

[48] Xiaofeng, Z. Yukun, Z. Beibei, W. Xing, C. and Binggen, R. (2002). Expression of the mouse metallothionein mutant ββ-cDNA in the lettuces (*Lactuca sativa* L.). *Chinese Science Bulletin*, 47 (7): 558-562.

[49] Goto, F. Yoshihara, T. and Saiki, H. (2000). Iron accumulation and enhanced growth in transgenic lettuce plants expressing the iron-binding protein ferritin. *Theoretical and Applied Genetics*, 100 (5): 658-664.

[50] Golczyk, H. Greiner, S. Wanner, G. Weihe, A. Bock, R. Börner, T. and Herrmann, R. G. (2014). Chloroplast DNA in mature and senescing leaves: a reappraisal. *The Plant Cell*, 26 (3): 847-854.

[51] Niki, T. Nishijima, T. Nakayama, M. Hisamatsu, T. Oyama-Okubo, N. Yamazaki, H. Hedden, P. Lange, T. Mander, L. N. and Koshioka, M. (2001). Production of dwarf lettuce by overexpressing a pumpkin gibberellin 20-oxidase gene. *Plant Physiology*, 126 (3): 965-972.

[52] Park, B.-J. Liu, Z. Kanno, A. and Kameya, T. (2005). Increased tolerance to salt-and water-deficit stress in transgenic lettuce (*Lactuca sativa* L.) by constitutive expression of LEA. *Plant Growth Regulation*, 45 (2): 165-171.

[53] Daniell, H. Cohill, P. R. Kumar, S. and Dufourmantel, N. (2004). Chloroplast genetic engineering, In *Molecular Biology and Biotechnology of Plant Organelles*, pp 443-490, Springer.

[54] Negrouk, V. Eisner, G. Lee, H.-i. Han, K. Taylor, D. and Wong, H. C. (2005). Highly efficient transient expression of functional recombinant antibodies in lettuce. *Plant Science*, 169 (2): 433-438.

[55] Joh, L. D. Wroblewski, T. Ewing, N. N. and VanderGheynst, J. S. (2005). High-level transient expression of recombinant protein in lettuce. *Biotechnology and Bioengineering*, 91 (7): 861-871.

[56] Beeramganti, N. Beeramganti, H. Subramanyam, K. and Rajasekhar, P. (2012). Chloroplast expression vector system & its transformation. *The International Journal of Science and Technology*, 2 (4): 1-12.

[57] Baecker, J. J. Sneddon, J. C. and Hollingsworth, M. J. (2009). Efficient translation in chloroplasts requires element (s) upstream of the putative ribosome binding site from atpI. *American Journal of Botany*, 96 (3): 627-636.

[58] Verma, D. and Daniell, H. (2007). Chloroplast vector systems for biotechnology applications. *Plant Physiology*, 145 (4): 1129-1143.

[59] Altpeter, F. Baisakh, N. Beachy, R. Bock, R. Capell, T. Christou, P. Daniell, H. Datta, K. Datta, S. and Dix, P. J. (2005). Particle bombardment and the genetic enhancement of crops: myths and realities. *Molecular Breeding*, 15 (3): 305-327.

[60] H Wani, S. Haider, N. Kumar, H. and B Singh, N. (2010). Plant plastid engineering. *Current Genomics*, 11 (7): 500-512.

[61] Scotti, N. Gargano, D. Lenzi, P. and Cardi, T. (2011). *Transformation of the plastid genome in higher plants*, Bentham Science Publishers Ltd.

[62] Daniell, H. Lin, C. S. Yu, M. and Chang, W. J. (2016). Chloroplast genomes: diversity, evolution, and applications in genetic engineering. *Genome Biology*, 17 (1): 134.

[63] Bock, R. (2001). Transgenic plastids in basic research and plant biotechnology. *Journal of Molecular Biology*, 312 (3): 425-438.

[64] Kanamoto, H. Yamashita, A. Asao, H. Okumura, S. Takase, H. Hattori, M. Yokota, A. and Tomizawa, K.-I. (2006). Efficient and stable transformation of *Lactuca sativa* L. cv. Cisco (lettuce) plastids. *Transgenic Research*, 15 (2): 205-217.

[65] Rigano, M. M. De Guzman, G. Walmsley, A. M. Frusciante, L. and Barone, A. (2013). Production of pharmaceutical proteins in solanaceae food crops. *International Journal of Molecular Sciences*, 14 (2): 2753-2773.

[66] Wani, S. H. Sah, S. K. Sagi, L. and Solymosi, K. (2015). Transplastomic plants for innovations in agriculture. *Agronomy for Sustainable Development*, 35 (4): 1391-1430.

[67] Daniell, H. Muthukumar, B. and Lee, S. (2001). Marker free transgenic plants: engineering the chloroplast genome without the use of antibiotic selection. *Current Genetics*, 39 (2): 109-116.

In: Lactuca: Cultivation and Uses
Editor: Jan Krüger
ISBN: 978-1-53617-729-9


*Chapter 5*

# *ULVA LACTUCA*: ADSORPTION, BIOACCUMULATION AND REMEDIATION

***María M. Areco[1,2,3], PhD, Vanesa N. Salomone[1,2,3], PhD and María dos Santos Afonso[2,*], PhD.***

[1]Instituto de Investigación e Ingeniería Ambiental (IIIA), UNSAM-CONICET, San Martín, Bs. As., Argentina
[2]INQUIMAE and Departamento de Química Inorgánica, Analítica y Química Física, Facultad de Ciencias Exactas y Naturales, Universidad de Buenos Aires, Buenos Aires, Argentina
[3]Consejo Nacional de Investigaciones Científicas y Técnicas. CONICET, Buenos Aires, Argentina

## ABSTRACT

Pollution of natural waters has become a major issue all over the world. As a result, scientists are studying new and alternative technologies not only to identify the presence of different pollutants in the environments but to remove them from waters and industrial effluents. In these senses, in recent years various red, green and brown

* Corresponding Author's E-mail: dosantos@qi.fcen.uba.ar.

seaweeds were investigated as potential bioindicators of contaminants in the environment; as well as their use in remediation process. The green marine macroalga, *Ulva lactuca* (Chlorophyta) is very common at coastal areas, and is a bloom forming macroalgae that occur in shallow estuaries. This specie has a high bonding affinity with heavy metals and several organic compounds due to the presence of different functional groups on its cell walls (such as carboxyl, hydroxyl, phosphate or amine) that can bind metal and/or organic ions. These characteristics makes it suitable to be used as a bioindicator of marine pollution and for remediation purposes. In general, the investigations are focused on the application of dried algae biomass to promote the removal of different contaminants, mainly metals from aqueous solutions (biosorption); however, the use of living organisms (bioaccumulation), may be advantageous. While growing, living macroalgae can remove simultaneously pollutants and nutrients (nitrates and phosphates) from wastewaters and capture $CO_2$ emissions. In this context, the present chapter will conduct a comprehensive inquiry on the advances related to the possibility of *Ulva lactuca* to be used as a bioindicator of contamination as well as its capability to be employed for the removal of pollutants from the surrounding environment. It will review these different approaches, and analyse the advantages and disadvantages of each as exposed on the literature, as well as the challenges that these alternatives confront as sustainable remediations technologies.

**Keywords**: *Ulva lactuca*, macroalgae, marine pollution, biomonitor, bioindicator

## INTRODUCTION

Environmental pollution in the marine coastline is an important topical issue in the context of ecological disturbance and climate change (Chakraborty et al., 2014). The presence of organic and inorganic toxics in coastal zones is a serious issue leading to a significant environmental and ecological degradation. The main contaminants found in marine environments are: potentially toxics metals; organic compounds, such as oil hydrocarbons, phenols, polycyclic aromatic hydrocarbons (PHAs) and organochlorine compounds, among others; and also the excess of nutrients such as nitrogen and phosphorous that causes eutrophication on many sea shores (Chen et al., 2018; Lewis Jr et al., 2011).

Trace metals are introduced into the coastal environments by various natural and anthropogenic activities such as the discharges from aquaculture, sewage, mining, gasoline from boats, agricultural, industrial and domestic effluents, accidental chemical spills, and dredging activity, related with rapid development, urbanization, and industrialization (Chakraborty et al., 2014; Satheeswaran et al., 2019). Metal pollution in aquatic ecosystems is a major environmental concern impacting various life-forms (Corales-Ultra et al., 2019), since they are highly persistent and can be toxic at low concentrations (Chaudhuri et al., 2007). In aquatic systems, metals are either dissolved in the water column or precipitated on the sediment bed, depending upon the nature of the chemical species and the physicochemical factors such as pH, conductivity, salinity and organic matter, which may exceed the permissible levels in sediment and seawater (Satheeswaran et al., 2019). However, analysis of total metal content in water and sediment does not predict the toxicity of contaminants to biota, because they can be accumulated in organisms of different trophic levels of the food chain, damaging their tissues and suppressing growth and ultimately affecting human beings. In this sense, elements potentially toxic can lead to various disorders, such as deactivate proteins, denature enzymes, and disturb cell functions (Bonanno et al., 2020) . The toxicity of heavy metals in the environment is related to their concentrations and environmental speciation and not to their essential characteristics. Heavy metals may be in waters as organic and inorganic colloids, mineral particles (suspended solids) and dissolved (cationic or anionic complexes in solution). Iron, cadmium and mercuric chloride stimulate the production of free reactive oxygen species, which in turn impair the functions of proteins, lipids and DNA causing oxidative stress and ultimately cell death (Chakraborty et al., 2014). Dietary exposure is a significant route for trace metals to the humans and constitutes about 90% of exposure (El-Kady and Abdel-Wahhab, 2018).

Regarding to organic pollution, gasoline is one of the most ubiquitous organic contaminants in the marine environment, often at a high level in areas submitted to intense ship traffic. The degradations reactions of organic contaminants with saturated bonds in their structure (mainly alkanes) are slow and require more energy; hence they are very resistant to degradation (Al-Hafedh et al., 2015). Accumulation of these substances can result highly toxic or mortal for aquatic organisms and can ultimately cause the decline of population and subsequent loss of marine ecosystem diversity and stability (El-Shoubaky and Mohammad, 2016). In addition, the attention for persistent organic pollutants (POPs) have increased due to widespread occurrence in the environments and their toxic, mutagenic and carcinogenic characteristics (Qiu et al., 2017).

Eutrophication due to an excess of nutrient is prevalent in shallow water ecosystems worldwide. The occurrence of green tides is frequently associated with anthropogenic nutrient inputs. *Ulva* is a genus of bloom forming macroalgae that occur in shallow estuaries worldwide and can create both ecological and economic problems (Smetacek and Zingone, 2013). *Ulva* sp. growth rates and tissue nitrogen (N) concentrations are often correlated with the seawater dissolved inorganic nitrogen concentrations (Fan et al., 2014); however, in areas with very high inorganic nitrogen concentrations, phosphorus and iron may limit macroalgal growth (Teichberg et al., 2010). In this sense, different Ulva species have being tested for nutrient remediation, especially in aquaculture systems, where integrate multi-trophic aquaculture, which involves farming of fed fish species together with "extractive" species such as seaweeds and filter feeders to take up inorganic and organic nutrients, respectively, has the potential to mitigate the environmental impacts of this kind of practices (Hadley et al., 2015).

In relation to the presence of pollutants in the marine environment, macroalgae may have different uses such as biomonitoring or bioremediation. Both processes are related to the characteristics of algae cell walls. The cellular walls of algae, principally its fibril matrix and intercellular spaces, are enriched with sulphated polysaccharides. Metals and organic pollutants are easily absorbed in cellular carbon and/or adsorbed to cell surfaces (Qiu et al., 2017). In this sense, many studies have evaluated the role of *Ulva lactuca* in relation with the presence of contaminants in the surrounding environment. This specie presents a promising set of characteristics for environmental applications; it has a cosmopolitan distribution, high growth rates, withstanding high variations of water salinity, high rates of nutrient assimilation, especially ammonium ($NH_4^+$), grows well in eutrophic waters and has tolerance to contaminants (Bonanno et al., 2020; Chen et al., 2015; Nielsen et al., 2012; Whitehouse and Lapointe, 2015). Moreover, it has a large surface area due to its thin and sheet-like thallus, and a relatively simple structure, with uniform and physiologically active cells that facilitates the interactions between the several functional groups in the surface and the metallic cations in solution (Sarı and Tuzen, 2008; Turner et al., 2007).

## Macroalgae and Environmental Contamination

Macroalgae constitute the base of trophic network in coastal rocky shores; these have been recognized as bioengineer organisms, which diversity and abundance determine the structure and complexity of intertidal and subtidal communities (Valdés et al., 2018). Contamination has been observed to be an important inducer of biological stress in macroalgae.

*Ulva lactuca*'s widespread distribution and its key ecological importance make it a target organism to address for potential responses to human-mediated environmental disturbances.

### Environmental Monitoring

Monitoring the environmental state of coastal habitats is of the utmost importance for the conservation of marine ecosystems and associated biodiversity (Bonanno et al., 2020). Bioindicators are organisms that contain information on quality of the environment (or a part of the environment) and biomonitors are organisms that contain information on the quantitative aspects of quality of the environment, providing a measure of the degree of the environmental contamination. For that, such organisms need to be sensitive to specific pollutants and tolerate high concentrations of these pollutants in the environment (Markert et al., 2003). Aquatic organisms are often used as both 'biomonitors' and 'bioindicators' of environmental pollution. There are requisites of an ideal biomonitor, some of which are: (i) the ability to accumulate contaminants without being severely affected or killed by encountered in the environment; (ii) sessile, in order to be representative of the study area; and (iii) abundant throughout most of the world's coasts (Phillips, 1980). Bioindicators and biomonitors of aquatic environments have been regularly used since the early 1990s (Rainbow and Phillips, 1993), algae are considered as good bioindicators and often used to assess marine ecosystems, especially for their high capacity of trace element bioaccumulation (Doshi et al., 2008). Furthermore, algae play an important role in the nutrient dynamics of coastal systems and reflect changes in water quality efficiently (Chaudhuri et al., 2007). Hence, many studies have been carried out across the globe using different species of algae including *U. lactuca*, as bioindicators of environmental contamination (Bonanno et al., 2020; Conti and Cecchetti, 2003; Mourad and Abd El-Azim, 2019).

### *Heavy Metals Contamination*

Metals are particularly significant for the ecotoxicology of the marine environment, since they are highly persistent and can be toxic in trace amounts. Heavy metals concentration in seawater is very low and fluctuates whenever the physicochemical parameters in the environment change. At the same time, heavy metal may accumulate in sediments but also the amount of metals adsorbed or precipitated in them are subjected to changes on sediment composition, grain size composition, pH and redox potential, among other physicochemical parameters. Since the analysis of total metal content in water and sediments does not predict the toxicity of contaminants to biota, the use of bioindicators allows also detecting the fraction of trace elements that has direct effects on living organisms (Bonanno et al., 2020). In this sense, organisms, such as macroalgae, can be used as monitors to give information of heavy metals concentrations in the surrounding environment, since macroalgae can accumulate trace metals, reaching concentration values that are thousands of times higher than the corresponding concentrations in sea water (Conti and Cecchetti, 2003). Nevertheless, the potential of living macroalgae and their bioaccumulation capabilities have been little explored despite its multiple advantage (Chojnacka, 2010; Henriques et al., 2017). While growing, living macroalgae can remove simultaneously contaminants (such as metals) and excess nutrients (nitrates and phosphates) from marine waters and capture $CO_2$ emissions (Henriques et al., 2017, 2015; Lartigue et al., 2003). Living macroalgae may lead to lower levels of pollutants in aquaculture wastewaters due to intracellular accumulation processes. Many authors indicate that bioaccumulation requires simpler installation compared with other treatment methods, as for example biosorption, since it is not necessary to apply pre-treatments, such as drying and activation of algal samples (Chojnacka, 2010). However, absorption mechanisms in living algae are more complex than in non-living algae since absorption takes place during the growth phase and intracellular uptake of heavy metal ions usually occur in this phase. In contrast, non-living algae cells absorb metal ions biosorption on the surface of the cell membrane and it is considered an extracellular process (Chojnacka, 2010).

Among the reasons why macroalgae are recognized as useful bioindicators to provide qualitative information of metal contamination level and environmental quality in seawater are found: their sedentary lifestyle, considerable biomass, and easy identification and collection (Chaudhuri et al., 2007). Trace elements are accumulated more intensively than those which play a major role in metabolism (Gherib et al., 2016). The accumulation of heavy metals in algae involves two processes: an initial rapid (passive) uptake followed by a much slower (active) uptake (Bates et al., 1982). During the passive uptake, metal ions adsorb onto the cell surface within a relatively short span of time (few seconds or minutes), and the process is metabolism independent. Active uptake is metabolism-dependent, causing the transport of metal ions across the cell membrane into the cytoplasm. In some instances, the transport of metal ions may also occur through passive diffusion owing to metal-induced increase in permeability of the cell membrane (Gadd, 1988).

### *Ulva Lactuca as Bioindicator of Heavy Metals*

*Ulva lactuca* have been recognized as one of the best bioindicators of marine pollution due to its notable capacity to accumulate trace elements under toxic conditions and mainly for being cosmopolitan and being on many coasts of the world (Bonanno et al., 2020; Chaudhuri et al., 2007; Conti and Cecchetti, 2003; Diop et al., 2016). A study carried out in different marine organism, demonstrates that algae are considered better biomonitors for metals in the water column than clams or other organisms with hard exoskeleton (El-Shoubaky and Mohammad, 2016).

*Ulva lactuca* has several attractive characteristics for water remediation. Its structure is simple, with a just two cell layers thick and sheet-like thallus, which translates into a large surface area, containing structurally uniform and physiologically active cells (Sarı and Tuzen, 2008). The cell wall is constituted by cellulose along with a high percentage of proteins bonded to polysaccharides (Romera et al., 2007; Trinelli et al., 2013), which comprises several functional groups such as amino, hydroxyl, carboxyl, and sulphate (Romera et al., 2007). This alga grows fast, since their photosynthetic products are quickly converted to cell growth. The macroalgal growth rate is closely linked to the removal efficiency, since a fast growth will lead to a rapid increase of biomass, and consequently to a multiplication of specific sites for metal binding (Chojnacka, 2010). It is perfectly adapted to the salinity variations occurring in estuaries due to tides and it withstands moderate eutrophic conditions. Additionally, high accumulation capabilities and tolerance to contamination of *U. lactuca* have been already reported. Studies demonstrate that metal bioaccumulation and toxic effects in important primary producers as *U. lactuca* may have implications for higher trophic levels (Jarvis and Bielmyer-Fraser, 2015).

The accumulation capacity of metals in algae depends on various factors: location, waves, exposure, temperature, salinity, photoperiod, pH, availability of nutrients, biomass age, the metabolism and the affinity of the species for each element (Kırbaşoğlu, 2001). It has been demonstrated that the main mechanisms of algae to thrive under metal excess are related to syntheses of metal chelators, the antioxidant metabolism, and cellular exclusion; in the latter, membrane active transport and chelation to the cell wall and epibionts play main roles (Moenne et al., 2016). Several investigations have demonstrated that cell wall chelation in different alga species can account for up to 95% of the total metal accumulated by the cell (Huang et al., 2013; Roncarati et al., 2015; Sáez et al., 2015, among others). Nevertheless, the cellular exclusion dynamics to direct metal exposure and associated development of intra- and inter-specific tolerance in green macroalgae, such as *Ulva lactuca*, has not being fully described, yet. Henriques et al. (2015) evaluated the uptake and accumulation of Hg using environmentally realistic concentrations, as well as the potential Hg methylation during removal was also addressed. The determination of total Hg content in algal biomass over time allowed to confirm and to follow the uptake of Hg by the living macroalgae and, to verify that volatilization of Hg or its conversion to organo-metallic forms was negligible during the decontamination process. Other researches assessed the capacity to removal in multi-metallic solutions, with excellent efficiencies (Henriques et al., 2017). Most of researchers follow an idealistic approach, assessing the performance of the removal process for mono-metallic solutions and the fact that live algae can simultaneously remove different metals is an interesting aspect because real waters are complex systems, containing a multiplicity of different ions, which interact and compete for binding sites, affecting removal efficiency. Henriques et al. (2019) exposed the capabilities of living *U. lactuca*, to remove As, Cd, Pb, Cu, Cr, Hg, Mn and Ni from contaminated waters. Their results demonstrated for the first time the capacity of living *U. lactuca* to remove simultaneously several top-priority hazardous substances (As, Pb, Hg and Cd) and other trace elements (Ni, Cr, Cu and Mn) and they support the hypothesis that macroalgae-based technologies may be a viable, cost-effective, and greener option to reduce the rejection of priority hazardous substances in contaminated waters. Besides, *U. lactuca* shows that the capacity to remove those elements from water was practically unaffected by salinity, which can interfere in the performance of conventional treatment procedures (Adeli et al., 2017).

**Table 1. Metal concentrations in *Ulva lactuca* from different world sites (mg kg$^{-1}$ dry weight)**

| Location | Cd | Cr | Cu | Ni | Pb | Zn | References |
|---|---|---|---|---|---|---|---|
| Coast of Sicily, Italy | 0.06–0.260 | 25–2.21 | 1.48–10.4 | 0.97–8.12 | 0.67–5.77 | 8.21–85.1 | (Bonanno et al., 2020) |
| Thermaikos, Greece | 0.42 | No data | 7.4 | 9.2 | 0.02 | 43.7 | (Sawidis et al., 2001) |
| Crete, Greece | 0.42-1.10 | No data | 7.00-14.5 | 8.7-13.7 | 0.02 | 16.4-56.3 | (Sawidis et al., 2001) |
| Apulian Coast, Italy | 0.20 ± 0.23 | No data | 12.1 ± 7.12 | No data | 0.84 ± 0.34 | 127 ± 60.2 | (Storelli et al., 2001) |
| Gulf of Gaeta, Italy | 0.18 ± 0.06 | 1.63 ± 0.60 | 5.8 ± 1.10 | No data | 1.94 ± 0.38 | 45.0 ± 12.0 | (Conti and Cecchetti, 2003) |
| El-Mex Bay, Egypt | 0.73 ± 0.24 | No data | 7.24 ± 2.3 | No data | No data | 27.4 ± 8.3 | (Abdallah and Abdallah, 2008) |
| Ouled Salh, coast of Honaine, Algeria | 0.15 ± 0.08 | No data | 0.45 ± 0.10 | 0.82 ± 0.17 | 1.44 ± 0.36 | 2.81 ± 0.54 | (Allam et al., 2016) |
| Eastern Harbour, Egypt | 1.84 ± 0.93 | No data | 14.5 ± 4.70 | No data | No data | 63.1 ± 3.20 | (Abdallah and Abdallah, 2008) |
| Tartous, Syria | 11.0 ± 0.33 | 4.71 ± 0.02 | 5.48 ± 0.28 | No data | 0.55 ± 0.99 | 11.0 ± 0.33 | (Al-Masri et al., 2003) |
| Abu-Qir Bay, Egypt | 1.27 | No data | 2.96 | 3.64 | 2.84 | 12.6 | (El-Din et al., 2014) |
| Suez Gulf, Egypt | 0.657 | 1.342 | 5.701 | 5.654 | 8.686 | 22.35 | (Mourad and Abd El-Azim, 2019) |
| Aqaba Gulf, Egypt | 0.531 | 1.419 | 5.33 | 4.285 | 8.112 | 13.52 | (Mourad and Abd El-Azim, 2019) |
| Suez Canal, Egypt | 0.593 | 1.276 | 8.915 | 4.479 | 7.186 | 13.9 | (Mourad and Abd El-Azim, 2019) |
| Valparaíso, Chile | 0.52 | No data | 10.7 | No data | No data | 13.0 | (Valdés et al., 2018) |
| Senegalese Coast | 0.30 ± 0.29 | 1.37 ± 1.16 | 6.23 ± 6.55 | 1.06 ± 0.32 | 2.64 ± 3.45 | 14.7 ± 15.1 | (Diop et al., 2016) |

Note: ranges are expressed with "–."

In these senses, many studies have evaluated the role of *U. lactuca* as a biomonitor of heavy metal pollution (such as Cd, Cr, Cu, Ni, Pb and Zn, among others), in different coast of the world with diverse levels of human impact. Table 1 resume many of the results obtained regarding metabolically metal uptake by *Ulva lactuca* from different world sites. Also, Bonanno et al. (2020), demonstrate that *U. lactuca* can accumulate essential and non-essential elements at similar concentrations and that the metal content in the biomass is significantly correlated with the levels of trace elements in water and sediments. Same results were obtained by Valdés et al. (2018). However, it is important to consider that patterns of metal concentrations in both sediments and *U. lactuca* not always follow the same trend within each site, since sediments are regularly washed away, and thus, metal concentrations in this matrix can change more rapidly than in algal biomass (Valdés et al., 2018); indeed, it has been fairly recognized that metal accumulation in macroalgae represent longer time frames of metal impacts on coastal environments (Brown and Depledge, 1998). The role played by chemical, physical and biological factors on trace element accumulation in natural environment by *U. lactuca* still need to be fully studied in order to completely understand the process.

Nevertheless, the capacity of *U. lactuca* to reflect the pollution levels due to trace elements in the marine environment seems to be proved, indicating that this is a suitable biomass for heavy metal biomonitoring studies.

### *Ulva Lactuca as Bioindicator of Persistent Organic Pollutants*

POPs attract increasing attention for their widespread occurrence in the environments and their toxic, mutagenic and carcinogenic characteristics. Algae, including phytoplankton and macroalgae are primary producers in aquatic ecosystems and play a key role in the transport of organic contaminants through the food chain to higher trophic levels, including humans. Algae have a high capacity to bind POPs, and then they are considered as suitable contaminant biomonitors. The extent of POP bioaccumulation by algae is related to the characteristics of organic substances, algae species, biomass, lipid content and environmental factors. The effect of growth dilution is another factor regulating the occurrence of chemicals in organisms. Algae growth rate is dependent on various factors, such as nutrient concentrations, temperature and light. Moreover biomass is not only controlled by algae growth, but also by grazed pressure (Qiu et al., 2017). On the other hand, many studies have reported that nitrogen deficiency induced a significant increase in algal lipid content and enhanced the bioconcentration of POPs (Chai et al., 2013; Cheng et al., 2014; Zhao et al., 2009); since POPs are hydrophobic, they tend to associate with lipids. Studies focused in bioconcentration and bioaccumulation of POPs, such as organochloride pesticides, in algae have increased in the last years (Leston et al., 2013; Qiu et al., 2017, among others).

Bioconcentration and bioaccumulation of POPs in macroalgae, including *U. lactuca* were not fully studied. Qiu et al. (2017) studied the occurrence of polybrominated diphenyl ethers (PBDEs) and organochlorine pesticides (OCPs) in phytoplankton and *U. lactuca*. Results demonstrate that the concentrations of both PBDEs and OCPs were an order of magnitude higher in phytoplankton than in *U. lactuca*, indicating that phytoplankton with larger surface areas has higher uptake efficiency for POPs than this macroalgae.

Leston et al. (2013) demonstrated that *U. lactuca* seems to be a good bioindicator of chloramphenicol (CAPh) an antibiotic, reflecting the concentration present in solution; and this ability points to a possibility of CAPh being accumulated and transferred along the trophic change through the consumption of *U. lactuca* by organisms in higher levels.

Gasoline may adhere to rocks, plants and animals, or penetrated into the sediments. Generally, it affects directly and indirectly aquatic organisms that live in the water body. A study exhibited that phenol and o-cresol, extracted from an oil refinery discharge, inhibited the germination of *Ulva lactuca* gametes at concentrations above 1 mg $L^{-1}$ (Jennings et al., 1996). Seaweeds and bivalves have been used as biomonitors of diesel fuel pollution (Nechev et al., 2002). However, a study on bioaccumulation of gasoline in different organism was carried out, it showed that accumulation level of *U. lactuca* was the more elevated (Al-Hafedh et al., 2015).

## MACROALGAE REMEDIATION POTENTIAL

The presence of a variety of organic and inorganic contaminants in wastewater is an environmental problem worldwide and its remediation has become a major issue in order to maintain adequate water quality for the environment and human health. As mentioned before, the increasing concentration of heavy metals is mainly due to industrial discharge; while the discharge of domestic effluents and agricultural activities are the main responsible for eutrophication, algae blooms, and in a great extent the presence of POPs in natural waters.

### Heavy Metal Biosorption

Anthropogenic and industrial-based activities release high amounts of heavy metals into aquatic environments, which significantly disturbs marine life due to their toxic and accumulative natures (Chakraborty et al., 2014; Corales-Ultra et al., 2019; Satheeswaran et al., 2019). Recently, heavy metal-related pollution in coastal regions has become a serious problem because of its significant environmental degradation (Chaudhuri et al., 2007). The impact of industrial discharges depends not only on their characteristics, such as biochemical oxygen demand, but also on the content of specific organic and inorganic substances. The characteristics of industrial wastewater may differ according to the type of industry which is involved. Heavy metals are present in many industrial effluents, examples of sources of heavy metal pollution are: tanneries whose waste with excess of Cu(II) and/or Cr(VI) go directly to water; the fuel and energy industries that generate million tons of As, Cd, Cr, Hg, Ni, Pb, Se and Zn per year; the metal industry that adds to the environment 0.39 million tonnes per year of some metals; while agriculture contributes with 1.4 million tonnes per year (Gadd and White, 1993). Consequently, industries are required to diminish the contents of heavy metals in their effluents to acceptable levels since their presence in the environment is a major threat to plants, animals and human health. The legislation and public concern for the environment have led to develop new techniques for the treatment of heavy metals present in effluents. Many techniques such as filtration, reverse osmosis, chemical precipitation, ion exchange, adsorption and plating have been used with varying success for the removal of toxic metals that contaminate aquatic environments. All these methods have advantages and disadvantages (Schiewer and Volesky, 1995). One of the main goals regarding heavy metals removal from waste waters consist in the reduction of these pollutants at very low levels (Lodeiro et al., 2006).

Biosorption, known as the sorption of heavy metals onto biological materials, is becoming a potential alternative for toxic metal removal from waters (Miretzky et al., 2006), since is an economical, efficient and sustainable method able to replace the most widely applied industrial materials such as activated carbon and ion-exchange resins by readily available biomass from nature, such as algae, fungus, bacteria or agricultural products. The prominent and emerging trend of subjecting biosorbent in the adsorption technology is mainly because of their natural existence, abundance, renewable, biodegradable and economic features. The adsorption isotherm equations used to describe the experimental data and the thermodynamic assumptions of the models often provide some insight into the sorption mechanism, the surface properties and affinity of the biosorbent (Rangabhashiyam et al., 2014).

The term "sorption" and the derivatives ad-, ab-, physico-, chemi- and bio- have their origin in the literal Latin verb "sorbere," a term used for the uptake of gaseous or solid substances dissolved without the exact knowledge of the mechanisms through which the process takes place (Valdman and Leite, 2000). The term adsorption refers to surface phenomena; the sequestration of the ions in solution may be based on physical phenomena or in a variety of chemical phenomena. Physical adsorption is not specific and the forces that bind the ions to the adsorbent are relatively weak. The activation energy for physical adsorption is generally not higher than 4.8 kJ mol$^{-1}$ (Volesky, 2007). Instead the chemical adsorption is specific and involves much stronger forces than those corresponding to physical adsorption, chemisorption examples are the processes of complexation and chelation. The term biosorption is difficult to define since is a physical process that includes a variety of mechanisms such as chemical adsorption, absorption, ion exchange, complexation and precipitation, many of these mechanisms may contribute to the overall process, depending on the sorbate, the biosorbent used, the environmental factors, and the presence or absence of metabolic processes in the case of living organisms. The prefix "bio" involves a biological entity, such as living organisms, or components or products produced or derived from a living organism, as in the case of biotechnology and bioengineering terms. Coupling "bio" to a physicochemical expression such as "sorption" denotes, then, that in the process is involved an organism, but not necessarily means that the process of sorption is different from sorption in abiotic systems (Gadd, 2009).

Specifically, biosorption is an innovative technology using living or dead biomasses to remove toxic metals and other pollutants from aqueous solutions. Various biomasses (e.g., bacteria, yeast, fungi, algae and agroindustry by-products and wastes (e.g., fruit peels, egg shells, crab shells and cork stoppers) with low economic value have been tested (Abdelfattah et al., 2016; Mohapatra et al., 2019; Silva et al., 2020).The prominent and emerging trend of subjecting biosorbent in the adsorption technology is mainly because of their natural existence, abundance, renewable, biodegradable and economic features. Among different biomass materials, the algae constitute an excellent biosorbent material because of high adsorption coefficients, the low-cost, available in large quantities, low sensitivity to environmental and impurity factors, and its excellent retention capacity, comparable in some cased with those of synthetic resins (Abbas et al., 2019; Silva et al., 2020).

A number of chemical groups that contribute to the biosorption of ions in solution have been reported, which includes hydroxyl, carboxyl, carbonyl, sulfide, amino, amide, phosphonate, phosphodiester, among others (Crist et al., 1981; Hunt, 1986). The importance of each of these groups in the adsorption of, for example heavy metals, depends not only on metal affinity with the group, but also on factors such as the number of adsorption sites on the adsorbent material, the accessibility to them and chemical form in which they occur, as was mentioned earlier in the present chapter. The main mechanism of metal cation adsorption is the formation of complexes between the metal ion and a functional group present on the surface or within the porous structure of the biological material (Fourest et al., 1996). A site can be occupied by a metal ion present in the solution when the bonding strength of the metal is greater than the ion which is previously adsorbed. The occupancy of the surface depends on the total number of surface sites, the free ion concentration in solution, the pH and temperature. Most binding sites in biosorption have acid-base properties (hydroxyl, carboxyl, sulfide, sulfonate, phosphonate), are neutral when they are protonated and negatively charged when deprotonated. When the pH of the solution exceeds the pKa of the groups, a larger number of binding sites are deprotonated and under these conditions the adsorption of cations is favoured. Ion exchange is the predominant mechanism in the biosorption of heavy metals by algal biomass and other types of natural biosorbents (Davis et al., 2003). This process is a reversible chemical reaction that takes place when an ion of a solution is exchanged for another of the same sign, which is attached to an immobile solid particle, so changes in the ionic strength of the solution modified the adsorption process. This process takes place continuously in nature, both in inorganic material as in living cells.

The functional groups on the sorbent surface are usually investigated by IR and FTIR spectroscopy. Scanning Electron Microscopy with X-ray microanalysis (SEM/EDS) are used to study the morphology of the adsorbent, before and after adsorption of metals, and provide quantitative analysis of elemental composition, SEM/EDS may also show the elemental distribution in a sample surface. Figures 1 A and B, show the superficial structure at two different magnifications (A: 1000X and B: 5000X) of *Ulva lactuca*, and Figure 2, A-E are the EDS analysis of the elemental composition of the algae surface, before and after the adsorption process of Cd(II), Cu(II), Zn(II) and Pb(II), respectively.

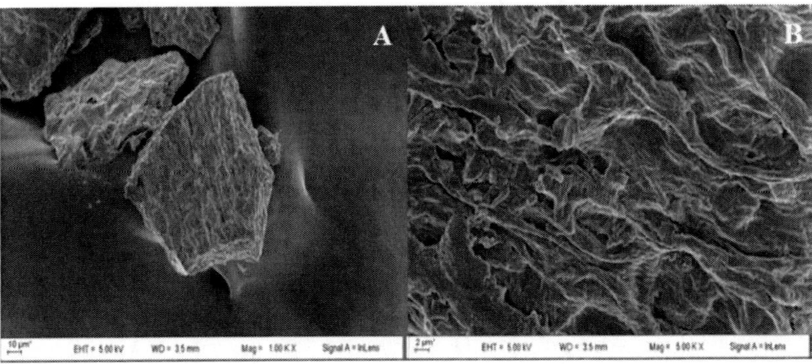


Figure 1. *U. lactuca* SEM images at two magnifications (A: 1000X y B:5000X) and 5 KV.

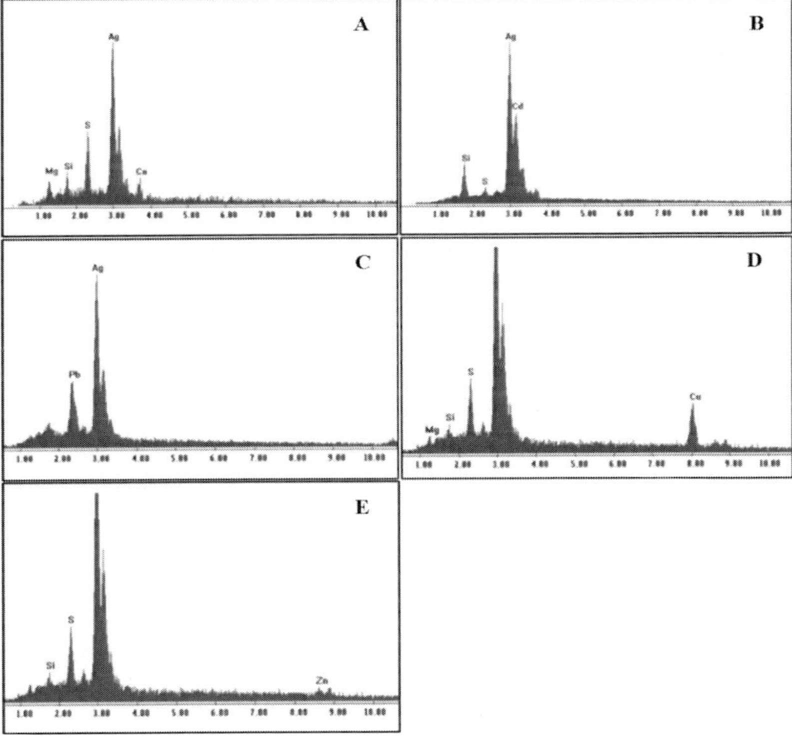


Figure 2. EDS images of *U. lactuca*: before treatment (A); and after Cd(II), Pb(II), Cu(II) and Zn(II) adsorption (B), (C), (D) and (E), respectively, at pH 5.5 and constant ionic strength.

The *Ulva* gender algae contain three polysaccharides: starch, cellulose and anionic compounds containing sulfate groups, the last two, are cell wall components. Chlorophyll a and b are also present as well as some carotenoids such as the β-carotene (Paradossi et al., 1999). The major sugars present in these algae are rhamnose, xylose, glucose and small amounts of mannose, galactose and arabinose (Pengzhan et al., 2003). The soluble fraction of the cellular wall is composed of a polysaccharide fraction described as a highly charged polyelectrolyte, named Ulvan (Figure 3), whose main component is a disaccharide the β-D-glucurono-syluronic acid (1→ 4) L-rhamnose-3-sulfate.

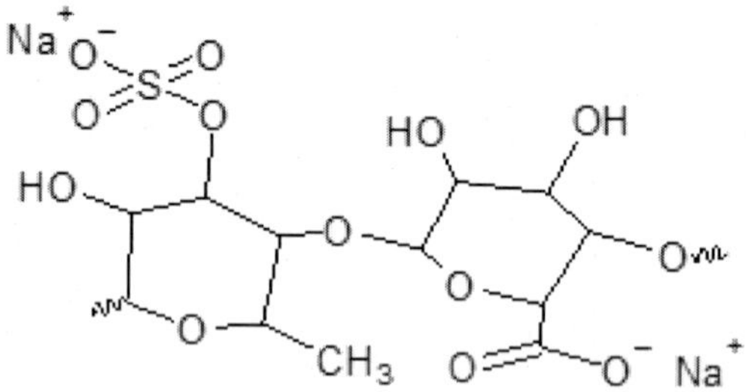

Figure 3. Structure of β-D-glucurono-syluronic acid (1→ 4) L-rhamnose-3-sulfate.

***U. Lactuca* Nitrogen and Phosphorous Remediation Potential**

In addition to the high capacity of *U. lactuca* to remove metals and other elements from water, they can be used as bioindicators of nutrients in the water column as their ability to assimilate surrounding nutrients is rapid which is clearly reflected by their tissue nutrient content within a relatively short period of time (Sutula, 2011). For that, there is a growing interest to combine the algae production to the aquaculture of fish and other marine invertebrates for consumption. Algaquaculture is a novel concept that integrates algae culture with aquaculture (Yang et al., 2020). Fed aquaculture (e.g., fish and shrimp) in coastal areas throughout the world has serious environmental impacts. Aquaculture waste, such as fish excretion and faeces, rich in inorganic nitrogen (N) and phosphorus (P), may significantly contribute to the nutrient loading of coastal waters leading to problems of considerable concern, including coastal eutrophication, loss of biodiversity and diseases (Diana et al., 2013). Integration of invertebrates or finfish aquaculture with seaweeds culture can be considered a practice for bioremediation of the nutrient laden effluents. The enhanced water quality in these systems leads to improved fish performance and higher biomass production, and to reduction in the energy power used, contributing to greater profitability (Cunha et al., 2019). Besides its ecological aspect, integrated aquaculture also has economic incentives as the nutrients contained in effluents, such as N and P, could be channelled into the production of valuable products, as algal biomass (Alemañ et al., 2019; Dominguez and Loret, 2019; Hurtado et al., 2019; Lubsch and Timmermans, 2018). Some researchers showed that *U. lactuca* can remove ammonia from water and sediments (Ho et al., 1999; Sode et al., 2013) and that nitrogen from anthropogenic sources contributed to its growth (Van Alstyne, 2016).

Besides the economic value of algae, seaweed farming can have positive environmental impacts because it makes use of nutrient emissions from fish farms and other anthropogenic nitrogen and phosphorous sources that enter the ocean and it can also take up anthropogenic carbon dioxide emissions that cause ocean acidification (Roleda and Hurd, 2019). The selection of seaweed species for their commercial use as biofilters depends of several aspects. First, is important the commercial value and, furthermore, on the physiological capabilities for rapid growth in culture conditions together with its capacity to accumulate of N and P to high levels in tissue (Roleda and Hurd, 2019). *U. lactuca* have a high growth rate and N and P uptake (Tremblay-Gratton et al., 2018). Seaweeds used for human consumption, biofuels and chemical extraction may be of relatively high economic value and can contribute substantially to the economic viability of integrated aquaculture systems (Buschmann and Camus, 2019). Species of the genus *Ulva* have been preferred in biofiltration studies and treatment of land-based pond/tank effluent due to a high biomass production and biofiltering efficiency (Shpigel et al., 2019). In particular, *U. lactuca* have been successful because of its large thin thalli that need more surfaces rather than the volume of culture vessel to absorb enough light for photosynthesis.

Numerous studies have focused on the integration of *Ulva lactuca* with fish, shrimp (Calheiros et al., 2019), abalone (Macchiavello and Bulboa, 2014), urchin (Shpigel et al., 2018) and mussel (Nardelli et al., 2019), demonstrating the benefits in size and growth rates on animals. An interesting aspect from the economic point of view is to find a use to the *Ulva* biomass obtained in the remediation process. In that sense, biomass of *U. lactuca* can be utilized for their nutritive value for animal feed e.g., flies, abalones and sea urchins which is another advantage of its cultivation (Dworjanyn et al., 2007; Kamio et al., 2016; Laramore et al., 2018; Qiu et al., 2017; Wan et al., 2019). A study demonstrated that *Ulva* cultivated under high nutrients offers a sustainable source of potential biomass for n-3 PUFA, pigments and phenolics with attributable anti-oxidant and anti-inflammatory activity (McCauley et al., 2018). Besides, *U. lactuca* has demonstrated potential for energy production (Bruhn et al., 2011). Pilot studies indicated that the yearly yield of *Ulva lactuca* is 4–5 times greater than terrestrial energy crops with a methane yield at a level between cow manure and energy crops (Bruhn et al., 2011).

## CONCLUSION

*Ulva lactuca* is a cosmopolitan alga; it has been widely studied around the world. Indeed, it adapts particularly well to highly eutrophic coastal environments and is therefore often used as a bioindicator species for the presence of urban and agricultural pollution. Studies show its great ability to accumulate toxic and trace elements and its role as environmental monitor in marine ecosystems. The bioaccumulation and biomagnification of metals in algae may affect trophic chain.

Besides, *U. lactuca* has been proven to be very efficient as a cost-effective wastewater remediator or/and for restoring water quality of the mariculture effluents based on the use of living macroalgae to promote removal of toxic metals and nutrients (e.g., nitrate, ammonia and phosphate) and to reduce the environmental impact derived from the high load of nutrients contained in these effluents. In addition, simultaneously *Ulva lactuca* could be cultivated successfully for biomass production for other uses and applications.

## REFERENCES

Abbas, M. N., Al-Hermizy, S. M. M., Abudi, Z. N., Ibrahim, T. A., 2019. Phenol biosorption from polluted aqueous solutions by Ulva *Lactuca* alga using batch mode unit. *J. Ecol. Eng.* 20, 225–235. https://doi.org/10.12911/22998993/109460.

Abdallah, M. A. M., Abdallah, A. M. A., 2008. Biomonitoring study of heavy metals in biota and sediments in the South Eastern coast of Mediterranean sea, *Egypt. Environ. Monit. Assess.* 146, 139–145.

Abdelfattah, I., Ismail, A. A., Sayed, F. Al, Almedolab, A., Aboelghait, K. M., 2016. Biosorption of heavy metals ions in real industrial wastewater using peanut husk as efficient and cost effective adsorbent. *Environ. Nanotechnology, Monit. Manag.* 6, 176–183. https://doi.org/10.1016/j.enmm.2016.10.007.

Adeli, M., Yamini, Y., Faraji, M., 2017. Removal of copper, nickel and zinc by sodium dodecyl sulphate coated magnetite nanoparticles from water and wastewater samples. *Arab. J. Chem.* 10, S514–S521. https://doi.org/10.1016/j.arabjc.2012.10.012.

Al-Hafedh, Y. S., Alam, A., Buschmann, A. H., 2015. Bioremediation potential, growth and biomass yield of the green seaweed, Ulva *Lactuca* in an integrated marine aquaculture system at the Red Sea coast of Saudi Arabia at different stocking densities and effluent flow rates. *Rev. Aquac.* 7, 161–171. https://doi.org/10.1111/raq.12060.

Al-Masri, M. S., Mamish, S., Budier, Y., 2003. Radionuclides and trace metals in eastern Mediterranean Sea algae. *J. Environ. Radioact.* 67, 157–168.

Alemañ, A. E., Robledo, D., Hayashi, L., 2019. Development of seaweed cultivation in Latin America: current trends and future prospects. *Phycologia* 58, 462–471.

Allam, H., Aouar, A., Benguedda, W., Bettioui, R., 2016. Use of sediment and algae for biomonitoring the Coast of Honaïne (far west Algerian). *Open J. Ecol.* 6, 159.

Bates, S. S., Tessier, A., Campbell, P. G. C., Buffle, J., 1982. Zinc adsorption and transport by chlamydomonas Varuiabilis and Scenedesmus subspicatus (chloropycheae) grown in semicontinuoes culture. *J. Phycol.* 18, 521–529.

Bonanno, G., Veneziano, V., Piccione, V., 2020. The alga *Ulva Lactuca* (Ulvaceae, Chlorophyta) as a bioindicator of trace element contamination along the coast of Sicily, Italy. *Sci. Total Environ.* https://doi.org/10.1016/j.scitotenv.2019.134329.

Brown, M. T., Depledge, M. H., 1998. Determinants of trace metal concentrations in marine organisms, in: *Metal Metabolism in Aquatic Environments.* Springer, pp. 185–217.

Bruhn, A., Dahl, J., Nielsen, H. B., Nikolaisen, L., Rasmussen, M. B., Markager, S., Olesen, B., Arias, C., Jensen, P. D., 2011. Bioenergy potential of *Ulva Lactuca*: Biomass yield, methane production and combustion. *Bioresour. Technol.* 102, 2595–2604. https://doi.org/10.1016/j.biortech.2010.10.010.

Buschmann, A. H., Camus, C., 2019. An introduction to farming and biomass utilisation of marine macroalgae. *Phycologia* 58, 443–445. https://doi.org/10.1080/00318884.2019.1638149.

Calheiros, A. C., Reis, R. P., Castelar, B., Cavalcanti, D. N., Teixeira, V. L., 2019. *Ulva* spp. as a natural source of phenylalanine and tryptophan to be used as anxiolytics in fish farming. *Aquaculture.*

Chai, C., Yin, X., Ge, W., Wang, J., 2013. Effects of nitrogen and phosphorus concentrations on the bioaccumulation of polybrominated diphenyl ethers by Prorocentrum donghaiense. *J. Environ. Sci.* 25, 376–385.

Chakraborty, S., Bhattacharya, T., Singh, G., Maity, J. P., 2014. Benthic macroalgae as biological indicators of heavy metal pollution in the marine environments: A biomonitoring approach for pollution assessment. *Ecotoxicol. Environ. Saf.* 100, 61–68. https://doi.org/10.1016/j.ecoenv.2013.12.003.

Chaudhuri, A., Mitra, M., Havrilla, C., Waguespack, Y., Schwarz, J., 2007. Heavy metal biomonitoring by seaweeds on the Delmarva Peninsula, east coast of the USA. *Bot. Mar.* 50, 151–158. https://doi.org/10.1515/BOT.2007.018.

Chen, B., Zou, D., Jiang, H., 2015. Elevated $CO_2$ exacerbates competition for growth and photosynthesis between Gracilaria lemaneiformis and Ulva *Lactuca*. *Aquaculture* 443, 49–55. https://doi.org/10.1016/j.aquaculture.2015.03.009.

Chen, M., Ding, S., Chen, X., Sun, Q., Fan, X., Lin, J., Ren, M., Yang, L., Zhang, C., 2018. Mechanisms driving phosphorus release during algal blooms based on hourly changes in iron and phosphorus concentrations in sediments. *Water Res.* 133, 153–164.

Cheng, P., Wang, J., Liu, T., 2014. Effects of nitrogen source and nitrogen supply model on the growth and hydrocarbon accumulation of immobilized biofilm cultivation of *B. braunii*. *Bioresour. Technol.* https://doi.org/10.1016/j.biortech.2014.05.045.

Chojnacka, K., 2010. Biosorption and bioaccumulation–the prospects for practical applications. *Environ. Int.* 36, 299–307.

Conti, M. E., Cecchetti, G., 2003. A biomonitoring study: trace metals in algae and molluscs from Tyrrhenian coastal areas. *Environ. Res.* 93, 99–112.

Corales-Ultra, O. G., Peja, R. P., Casas, E. V., 2019. Baseline study on the levels of heavy metals in seawater and macroalgae near an abandoned mine in Manicani, Guiuan, Eastern Samar, Philippines. *Mar. Pollut. Bull.* 149, 110549. https://doi.org/10.1016/j.marpolbul.2019.110549.

Crist, R. H., Oberholser, K., Shank, N., Nguyen, M., 1981. Nature of bonding between metallic ions and algal cell walls. *Environ. Sci. Technol.* 15, 1212–1217.

Cunha, M. E., Quental-Ferreira, H., Parejo, A., Gamito, S., Ribeiro, L., Moreira, M., Monteiro, I., Soares, F., Pousão-Ferreira, P., 2019. Understanding the individual role of fish, oyster, phytoplankton and macroalgae in the ecology of integrated production in earthen ponds. *Aquaculture* 512, 734297. https://doi.org/10.1016/j.aquaculture.2019.734297.

Davis, T. A., Volesky, B., Mucci, A., 2003. A review of the biochemistry of heavy metal biosorption by brown algae. *Water Res.* 37, 4311–4330.

Diana, J. S., Egna, H. S., Chopin, T., Peterson, M. S., Cao, L., Pomeroy, R., Verdegem, M., Slack, W. T., Bondad-Reantaso, M. G., Cabello, F., 2013. Responsible Aquaculture in 2050: Valuing Local Conditions and Human Innovations Will Be Key to Success. *Bioscience* 63, 255–262. https://doi.org/10.1525/bio.2013.63.4.5.

Diop, M., Howsam, M., Diop, C., Goossens, J. F., Diouf, A., Amara, R., 2016. Assessment of trace element contamination and bioaccumulation in algae (*Ulva Lactuca*), mussels (*Perna perna*), shrimp (*Penaeus kerathurus*), and fish (*Mugil cephalus*, *Saratherondon melanotheron*) along the Senegalese coast. *Mar. Pollut. Bull.* 103, 339–343.

Dominguez, H., Loret, E. P., 2019. *Ulva Lactuca*, A Source of Troubles and Potential Riches. *Mar. Drugs* 17, 357.

Doshi, H., Seth, C., Ray, A., Kothari, I. L., 2008. Bioaccumulation of heavy metals by green algae. *Curr. Microbiol.* 56, 246–255.

Dworjanyn, S. A., Pirozzi, I., Liu, W., 2007. The effect of the addition of algae feeding stimulants to artificial diets for the sea urchin *Tripneustes gratilla*. *Aquaculture* 273, 624–633. https://doi.org/10.1016/j.aquaculture.2007.08.023.

El-Din, N. G. S., Mohamedein, L. I., El-Moselhy, K. M., 2014. Seaweeds as bioindicators of heavy metals off a hot spot area on the Egyptian Mediterranean Coast during 2008–2010. *Environ. Monit. Assess.* 186, 5865–5881.

El-Kady, A. A., Abdel-Wahhab, M. A., 2018. Occurrence of trace metals in foodstuffs and their health impact. *Trends Food Sci. Technol.* 75, 36–45.

El-Shoubaky, G. A., Mohammad, S. H., 2016. Bioaccumulation of gasoline in brackish green algae and popular clams. *Egypt. J. Aquat. Res.* 42, 91–98. https://doi.org/10.1016/j.ejar.2015.07.005.

Fan, X., Xu, D., Wang, Y., Zhang, X., Cao, S., Mou, S., Ye, N., 2014. The effect of nutrient concentrations, nutrient ratios and temperature on photosynthesis and nutrient uptake by *Ulva prolifera*: implications for the explosion in green tides. *J. Appl. Phycol.* 26, 537–544.

Fourest, E., Serre, A., Roux, J., 1996. Contribution of carboxyl groups to heavy metal binding sites in fungal wall. *Toxicol. Environ. Chem.* 54, 1–10.

Gadd, G. M., 2009. Biosorption: critical review of scientific rationale, environmental importance and significance for pollution treatment. *J. Chem. Technol. Biotechnol. Int. Res. Process. Environ. Clean Technol.* 84, 13–28.

Gadd, G. M., 1988. Accumulation of metals by microorganisms and algae. *Biotechnol.* Vol. 6b, Spec. Microb. Process. 401–433.

Gadd, G. M., White, C., 1993. Microbial treatment of metal pollution—a working biotechnology? *Trends Biotechnol.* 11, 353–359.

Gherib, A., Bedouh, Y., Messai, K., Menad, A., 2016. Biomonitoring of maritime pollution by heavy metals (Pb and Zn) in the coast of Jijel (Algeria). *Int. J. Environ. Stud.* 73, 214–225. https://doi.org/10.1080/00207233.2016.1143699.

Hadley, S., Wild-Allen, K., Johnson, C., Macleod, C., 2015. Modeling macroalgae growth and nutrient dynamics for integrated multi-trophic aquaculture. *J. Appl. Phycol.* 27, 901–916.

Henriques, B., Rocha, L. S., Lopes, C. B., Figueira, P., Duarte, A. C., Vale, C., Pardal, M. A., Pereira, E., 2017. A macroalgae-based biotechnology for water remediation: simultaneous removal of Cd, Pb and Hg by living *Ulva Lactuca*. *J. Environ. Manage.* 191, 275–289.

Henriques, B., Rocha, L. S., Lopes, C. B., Figueira, P., Monteiro, R. J. R., Duarte, A. C., Pardal, M. A., Pereira, E., 2015. Study on bioaccumulation and biosorption of mercury by living marine macroalgae: prospecting for a new remediation biotechnology applied to saline waters. *Chem. Eng. J.* 281, 759–770.

Ho, K. T., Kuhn, A., Pelletier, M. C., Burgess, R. M., Helmstetter, A., 1999. Use of *Ulva Lactuca* to distinguish pH-dependent toxicants in marine waters and sediments. *Environ. Toxicol. Chem.* 18, 207–212.

https://doi.org/10.1897/1551-5028(1999)018<0207:UOULTD>2.3.CO;2.

Huang, H., Liang, J., Wu, X., Zhang, H., Li, Q., Zhang, Q., 2013. Comparison in copper accumulation and physiological responses of *Gracilaria lemaneiformis* and *G. lichenoides* (Rhodophyceae). *Chinese J. Oceanol. Limnol.* 31, 803–812.

Hunt, S., 1986. Diversity of biopolymer structure and its potential for ion-binding applications. *Immobil. ions by bio-sorption.* Ellis Horwood Ltd., West Sussex, United Kingdom 15–46.

Hurtado, A. Q., Neish, I. C., Critchley, A. T., 2019. Phyconomy: the extensive cultivation of seaweeds, their sustainability and economic value, with particular reference to important lessons to be learned and transferred from the practice of eucheumatoid farming. *Phycologia* 58, 472–483. https://doi.org/10.1080/00318884.2019.1625632.

Jarvis, T. A., Bielmyer-Fraser, G. K., 2015. Accumulation and effects of metal mixtures in two seaweed species. *Comp. Biochem. Physiol. Part - C Toxicol. Pharmacol.* 171, 28–33. https://doi.org/10.1016/j.cbpc.2015.03.005.

Jennings, J. G., De Nys, R., Charlton, T. S., Duncan, M. W., Steinberg, P. D., 1996. Phenolic compounds in the nearshore waters of Sydney, Australia. *Mar. Freshw. Res.* 47, 951–959. https://doi.org/10.1071/MF9960951.

Kamio, M., Koyama, M., Hayashihara, N., Hiei, K., Uchida, H., Watanabe, R., Suzuki, T., Nagai, H., 2016. Sequestration of Dimethyl sulfoniopropionate (DMSP) and Acrylate from the Green Alga *Ulva* Spp. by the Sea Hare *Aplysia juliana*. *J. Chem. Ecol.* 42, 452–460. https://doi.org/10.1007/s10886-016-0703-1.

Kırbaşoğlu, Ç., 2001. Heavy Metals in Marine Algae from Şile in the Black Sea, 1994–1997. *Bull. Environ. Contam. Toxicol.* 67, 0288–0294. https://doi.org/10.1007/s00128-001-0123-x.

Laramore, S., Baptiste, R., Wills, P. S., Hanisak, M. D., 2018. Utilization of IMTA-produced Ulva *Lactuca* to supplement or partially replace pelleted diets in shrimp (*Litopenaeus vannamei*) reared in a clear water production system. *J. Appl. Phycol.* 30, 3603–3610. https://doi.org/10.1007/s10811-018-1485-3.

Lartigue, J., Neill, A., Hayden, B. L., Pulfer, J., Cebrian, J., 2003. The impact of salinity fluctuations on net oxygen production and inorganic nitrogen uptake by *Ulva Lactuca* (Chlorophyceae). *Aquat. Bot.* 75, 339–350. https://doi.org/10.1016/S0304-3770(02)00193-6.

Leston, S., Nunes, M., Viegas, I., Ramos, F., Pardal, M. Â., 2013. The effects of chloramphenicol on *Ulva Lactuca*. *Chemosphere* 91, 552–557. https://doi.org/10.1016/j.chemosphere.2012.12.061.

Lewis Jr, W. M., Wurtsbaugh, W. A., Paerl, H. W., 2011. Rationale for control of anthropogenic nitrogen and phosphorus to reduce eutrophication of inland waters. *Environ. Sci. Technol.* 45, 10300–10305.

Lodeiro, P., Barriada, J. L., Herrero, R., De Vicente, M. E. S., 2006. The marine macroalga *Cystoseira baccata* as biosorbent for cadmium (II) and lead (II) removal: kinetic and equilibrium studies. *Environ. Pollut.* 142, 264–273.

Lubsch, A., Timmermans, K., 2018. Uptake kinetics and storage capacity of dissolved inorganic phosphorus and corresponding N:P dynamics in *Ulva Lactuca* (Chlorophyta). *J. Phycol.* 54, 215–223. https://doi.org/10.1111/jpy.12612.

Macchiavello, J., Bulboa, C., 2014. Nutrient uptake efficiency of *Gracilaria chilensis* and *Ulva Lactuca* in an IMTA system with the red abalone *Haliotis rufescens*. *Lat. Am. J. Aquat. Res.* 42, 523–533. https://doi.org/10.3856/vol42-issue3-fulltext-12.

Markert, B. A., Breure, A. M., Zechmeister, H. G., 2003. Definitions, strategies and principles for bioindication/biomonitoring of the environment, in: *Trace Metals and Other Contaminants in the Environment*. Elsevier, pp. 3–39.

McCauley, J. I., Winberg, P. C., Meyer, B. J., Skropeta, D., 2018. Effects of nutrients and processing on the nutritionally important metabolites of *Ulva* sp. (Chlorophyta). *Algal Res.* 35, 586–594. https://doi.org/10.1016/j.algal.2018.09.016.

Miretzky, P., Saralegui, A., Cirelli, A. F., 2006. Simultaneous heavy metal removal mechanism by dead macrophytes. *Chemosphere* 62, 247–254.

Moenne, A., González, A., Sáez, C. A., 2016. Mechanisms of metal tolerance in marine macroalgae, with emphasis on copper tolerance in Chlorophyta and Rhodophyta. *Aquat. Toxicol.* 176, 30–37.

Mohapatra, R. K., Parhi, P. K., Pandey, S., Bindhani, B. K., Thatoi, H., Panda, C. R., 2019. Active and passive biosorption of Pb(II)using live and dead biomass of marine bacterium Bacillus xiamenensis PbRPSD202: Kinetics and isotherm studies. *J. Environ. Manage.* 247, 121–134. https://doi.org/10.1016/j.jenvman.2019.06.073.

Mourad, F. A., Abd El-Azim, H., 2019. Use of green alga Ulva *Lactuca* (L.) as an indicator to heavy metal pollution at intertidal waters in Suez Gulf, Aqaba Gulf and Suez Canal, Egypt. *Egypt. J. Aquat. Biol. Fish.* 23, 437–449.

Nardelli, A. E., Chiozzini, V. G., Braga, E. S., Chow, F., 2019. Integrated multi-trophic farming system between the green seaweed *Ulva Lactuca*, mussel, and fish: a production and bioremediation solution. *J. Appl. Phycol.* 31, 847–856. https://doi.org/10.1007/s10811-018-1581-4.

Nechev, J. T., Khotimchenko, S. V, Ivanova, A. P., Stefanov, K. L., Dimitrova-Konaklieva, S. D., Andreev, S., Popov, S. S., 2002. Effect of diesel fuel pollution on the lipid composition of some wide-spread Black Sea algae and invertebrates. *Zeitschrift für Naturforsch.* C 57, 1–5.

Nielsen, M. M., Bruhn, A., Rasmussen, M. B., Olesen, B., Larsen, M. M., Møller, H. B., 2012. Cultivation of *Ulva Lactuca* with manure for simultaneous bioremediation and biomass production. *J. Appl. Phycol.* 24, 449–458. https://doi.org/10.1007/s10811-011-9767-z.

Paradossi, G., Cavalieri, F., Pizzoferrato, L., Liquori, A. M., 1999. A physico-chemical study on the polysaccharide ulvan from hot water extraction of the macroalga *Ulva. Int. J. Biol. Macromol.* 25, 309–315.

Pengzhan, Y., Quanbin, Z., Ning, L., Zuhong, X., Yanmei, W., Zhi'en, L., 2003. Polysaccharides from *Ulva pertusa* (Chlorophyta) and preliminary studies on their antihyperlipidemia activity. *J. Appl. Phycol.* 15, 21–27.

Phillips, D. J. H., 1980. Quantitative aquatic biological indicators: their use to monitor trace metal and organochlorine pollution.

Qiu, S., Ge, S., Champagne, P., Robertson, R. M., 2017. Potential of *Ulva Lactuca* for municipal wastewater bioremediation and fly food. *Desalin. Water Treat.* 91, 23–30. https://doi.org/10.5004/dwt.2017.20767.

Qiu, Y. W., Zeng, E. Y., Qiu, H., Yu, K., Cai, S., 2017. Bioconcentration of polybrominated diphenyl ethers and organochlorine pesticides in algae is an important contaminant route to higher trophic levels. *Sci. Total Environ.* 579, 1885–1893. https://doi.org/10.1016/j.scitotenv.2016.11.192.

Rainbow, P. S., Phillips, D. J. H., 1993. Cosmopolitan biomonitors of trace metals. *Mar. Pollut. Bull.* 26, 593–601.

Rangabhashiyam, S., Anu, N., Nandagopal, M. S. G., Selvaraju, N., 2014. Relevance of isotherm models in biosorption of pollutants by agricultural byproducts. *J. Environ. Chem. Eng.* 2, 398–414.

Roleda, M. Y., Hurd, C. L., 2019. Seaweed nutrient physiology: application of concepts to aquaculture and bioremediation. *Phycologia* 58, 552–562. https://doi.org/10.1080/00318884.2019.1622920.

Romera, E., González, F., Ballester, A., Blázquez, M. L., Muñoz, J. A., 2007. Comparative study of biosorption of heavy metals using different types of algae. *Bioresour. Technol.* 98, 3344–3353. https://doi.org/10.1016/j.biortech.2006.09.026.

Roncarati, F., Sáez, C. A., Greco, M., Gledhill, M., Bitonti, M. B., Brown, M. T., 2015. Response differences between Ectocarpus siliculosus populations to copper stress involve cellular exclusion and induction of the phytochelatin biosynthetic pathway. *Aquat. Toxicol.* 159, 167–175.

Sáez, C. A., Roncarati, F., Moenne, A., Moody, A. J., Brown, M. T., 2015. Copper-induced intra-specific oxidative damage and antioxidant responses in strains of the brown alga Ectocarpus siliculosus with different pollution histories. *Aquat. Toxicol.* 159, 81–89.

Sarı, A., Tuzen, M., 2008. Biosorption of Pb (II) and Cd (II) from aqueous solution using green alga (*Ulva Lactuca*) biomass. *J. Hazard. Mater.* 152, 302–308.

Satheeswaran, T., Yuvaraj, P., Damotharan, P., Karthikeyan, V., Jha, D. K., Dharani, G., Balasubramanian, T., Kirubagaran, R., 2019. Assessment of trace metal contamination in the marine sediment, seawater, and bivalves of Parangipettai, southeast coast of India. *Mar. Pollut. Bull.* 149, 110499. https://doi.org/10.1016/j.marpolbul.2019.110499.

Sawidis, T., Brown, M. T., Zachariadis, G., Sratis, I., 2001. Trace metal concentrations in marine macroalgae from different biotopes in the Aegean Sea. *Environ. Int.* 27, 43–47.

Schiewer, S., Volesky, B., 1995. Modeling of the proton-metal ion exchange in biosorption. *Environ. Sci. Technol.* 29, 3049–3058.

Shpigel, M., Guttman, L., Ben-Ezra, D., Yu, J., Chen, S., 2019. Is Ulva sp. able to be an efficient biofilter for mariculture effluents? *J. Appl. Phycol.* 31, 2449–2459. https://doi.org/10.1007/s10811-019-1748-7.

Shpigel, M., Shauli, L., Odintsov, V., Ben-Ezra, D., Neori, A., Guttman, L., 2018. The sea urchin, Paracentrotus lividus, in an Integrated Multi-Trophic Aquaculture (IMTA) system with fish (*Sparus aurata*) and seaweed (*Ulva Lactuca*): Nitrogen partitioning and proportional configurations. *Aquaculture* 490, 260–269. https://doi.org/10.1016/j.aquaculture.2018.02.051.

Silva, B., Martins, M., Rosca, M., Rocha, V., Lago, A., Neves, I.C., Tavares, T., 2020. Waste-based biosorbents as cost-effective alternatives to commercial adsorbents for the retention of fluoxetine from water. *Sep. Purif. Technol.* 235, 116139. https://doi.org/10.1016/j.seppur.2019.116139.

Smetacek, V., Zingone, A., 2013. Green and golden seaweed tides on the rise. *Nature* 504, 84–88.

Sode, S., Bruhn, A., Balsby, T. J. S., Larsen, M. M., Gotfredsen, A., Rasmussen, M. B., 2013. Bioremediation of reject water from anaerobically digested waste water sludge with macroalgae (*Ulva*

*Lactuca*, Chlorophyta). *Bioresour. Technol.* 146, 426–435. https://doi.org/10.1016/j.biortech.2013.06.062.

Storelli, M. M., Storelli, A., Marcotrigiano, G. O., 2001. Heavy metals in the aquatic environment of the Southern Adriatic Sea, Italy: macroalgae, sediments and benthic species. *Environ. Int.* 26, 505–509.

Sutula, M., 2011. *Review of Indicators for Development of Nutrient Numeric Endpoints in California Estuaries* 269.

Teichberg, M., Fox, S. E., Olsen, Y. S., Valiela, I., Martinetto, P., Iribarne, O., Muto, E. Y., Petti, M. A. V, Corbisier, T. N., Soto-Jimenez, M., 2010. Eutrophication and macroalgal blooms in temperate and tropical coastal waters: nutrient enrichment experiments with *Ulva* spp. *Glob. Chang. Biol.* 16, 2624–2637.

Tremblay-Gratton, A., Boussin, J. C., Tamigneaux, Vandenberg, G. W., Le François, N. R., 2018. Bioremediation efficiency of Palmaria palmata and *Ulva Lactuca* for use in a fully recirculated cold-seawater naturalistic exhibit: effect of high NO3 and PO4 concentrations and temperature on growth and nutrient uptake. *J. Appl. Phycol.* 30, 1295–1304. https://doi.org/10.1007/s10811-017-1333-x.

Trinelli, M. A., Areco, M. M., Dos Santos Afonso, M., 2013. Co-biosorption of copper and glyphosate by *Ulva Lactuca*. *Colloids Surfaces B Biointerfaces* 105. https://doi.org/10.1016/j.colsurfb.2012.12.047.

Turner, A., Lewis, M. S., Shams, L., Brown, M. T., 2007. Uptake of platinum group elements by the marine macroalga, *Ulva Lactuca*. *Mar. Chem.* 105, 271–280.

Valdés, F. A., Lobos, M. G., Díaz, P., Sáez, C. A., 2018. Metal assessment and cellular accumulation dynamics in the green macroalga *Ulva Lactuca*. *J. Appl. Phycol.* 30, 663–671.

Valdman, E., Leite, S. G. F., 2000. Biosorption of Cd, Zn and Cu by *Sargassum* sp. waste biomass. *Bioprocess Eng.* 22, 171–173.

Van Alstyne, K. L., 2016. Seasonal changes in nutrient limitation and nitrate sources in the green macroalga *Ulva Lactuca* at sites with and without green tides in a northeastern Pacific embayment. *Mar. Pollut. Bull.* 103, 186–194. https://doi.org/10.1016/j.marpolbul.2015.12.020.

Volesky, B., 2007. Biosorption and me. *Water Res.* 41, 4017–4029.

Wan, A. H. L., Davies, S. J., Soler-Vila, A., Fitzgerald, R., Johnson, M. P., 2019. Macroalgae as a sustainable aquafeed ingredient. *Rev. Aquac.* 11, 458–492. https://doi.org/10.1111/raq.12241.

Whitehouse, L. N. A., Lapointe, B. E., 2015. Comparative ecophysiology of bloom-forming macroalgae in the Indian River Lagoon, Florida: *Ulva Lactuca*, Hypnea musciformis, and *Gracilaria tikvahiae. J. Exp. Mar. Bio. Ecol.* 471, 208–216. https://doi.org/10.1016/j.jembe.2015.06.012.

Yang, L., Wang, R., Lu, Q., Liu, H., 2020. "Algaquaculture" integrating algae-culture with aquaculture for sustainable development. *J. Clean. Prod.* 244. https://doi.org/10.1016/j.jclepro.2019.118765.

Zhao, Y., Yu, Z., Song, X., Cao, X., 2009. Biochemical compositions of two dominant bloom-forming species isolated from the Yangtze River Estuary in response to different nutrient conditions. *J. Exp. Mar. Bio. Ecol.* 368, 30–36.

In: Lactuca: Cultivation and Uses
Editor: Jan Krüger
ISBN: 978-1-53617-729-9


*Chapter 6*

# PRODUCTION OF VITAMIN $B_{12}$-ENRICHED LETTUCE FOR VEGETARIANS AND ELDERLY PEOPLE

***Tomohiro Bito and Fumio Watanabe****
Department of Agricultural, Life and Environmental Sciences,
Faculty of Agriculture, Tottori University, Tottori, Japan

## ABSTRACT

Lettuce (*Lactuca sativa* L.) is a popular leafy vegetable that can be easily cultivated and is a good source of various nutrients. However, plants, including vegetables, generally contain no or trace amounts of vitamin $B_{12}$ as they do not require this vitamin $B_{12}$ for growth. Therefore, if a sufficient amount of free vitamin $B_{12}$ can be incorporated into lettuce leaves grown under hydroponic conditions, it would become an excellent source of free vitamin $B_{12}$ for vegetarians and elderly people. Lettuce leaves were grown for 30 days under hydroponic culture conditions and treated with 5 μM vitamin $B_{12}$ for 24 h, after which the vitamin content in the leaves increased significantly from nondetectable levels to 164.6 ± 74.7 ng/g fresh weight. Thus, the recommended dietary allowance of

* Corresponding Author's E-mail: watanabe@tottori-u.ac.jp.

vitamin $B_{12}$ for adults (2.4 μg/day) in the United States and Japan can be met by consuming only two or three of these fresh leaves. Moreover, the majority (86%) of vitamin $B_{12}$ in these leaves was recovered in the form of free vitamin $B_{12}$ fractions. These results indicate that vitamin $B_{12}$-enriched lettuce leaves could be used as an excellent source of free vitamin $B_{12}$, particularly in elderly people with food-bound vitamin $B_{12}$ malabsorption.

**Keywords**: hydroponic culture, lettuce, *Lactuca sativa* L., vitamin $B_{12}$

## Introduction

Vitamin $B_{12}$ ($B_{12}$) (molecular weight = 1355.4) belongs to the "corrinoids" group, which contains a corrin macrocycle (Figure 1). The term "vitamin $B_{12}$" is generally restricted to cyanocobalamin (CN-$B_{12}$), which is the most chemically stable form of $B_{12}$ [1]. The recommended dietary allowance (RDA) of $B_{12}$ for adults is 2.4 μg/day in the United States and Japan [2]. $B_{12}$ is readily converted into two coenzyme forms of $B_{12}$ in living cells, namely, 5′-deoxyadenosyl-$B_{12}$ (Ado-$B_{12}$) and methyl-$B_{12}$ ($CH_3$-$B_{12}$), which function as coenzymes for methylmalonyl-CoA mutase (EC 5.4.99.2) [3] and methionine synthase (EC 2.1.1.13) [4], respectively. Methylmalonyl-CoA mutase catalyzes the conversion of *R*-methylmalonyl-CoA into succinyl-CoA through the citric acid cycle in the catabolic pathways of branched-chain amino acids and odd-numbered fatty acids (Figure 2A). Methionine synthase catalyzes the biosynthetic reaction of methionine from homocysteine and $N^{5'}$-methyltetrahydrofolate (Figure 2B). Deficiency of $B_{12}$ leads to the excess accumulation of methylmalonic acid, which potently inhibits the activity of succinate dehydrogenase in the citric acid cycle and further leads to the disruption of normal glucose and glutamic acid metabolism [5]. Futhermore, $B_{12}$ deficiency reportedly results in significant increases in homocysteine levels, which induces oxidative damage in various cellular components. Therefore, $B_{12}$ deficiency can induce various diseases such as megaloblastic anemia, developmental disorders, growth retardation, and neuropathy [2].

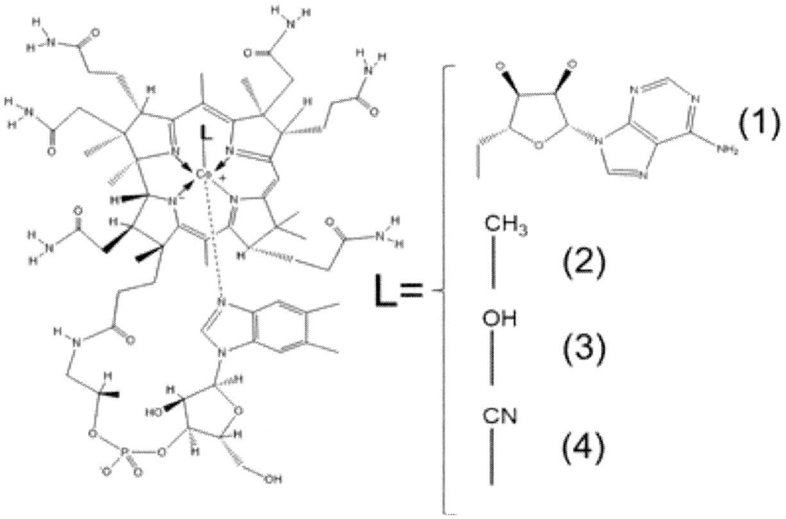

Figure 1. Structural formula of vitamin $B_{12}$ and partial structures of vitamin $B_{12}$-related compounds. (1) 5′-Deoxyadenosylcobalamin, (2) Methylcobalamin, (3) Hydroxocobalamin, and (4) Cyanocobalamin (or vitamin $B_{12}$).

$B_{12}$ is the only well-known vitamin that is synthesized only by certain bacteria and archaeon. $B_{12}$ is primarily concentrated in the bodies of predators located at a higher level in the food chain [6]. Therefore, foods (meat, milk, eggs, fish, and shellfish) derived from animals are the major dietary sources of $B_{12}$ [7]. Due to the absence of $B_{12}$ in plant-derived food sources, vegetarians face a higher risk of $B_{12}$ deficiency, especially pregnant women, children, adolescents, and elderly peopole [8, 9].

A large number of people have low serum $B_{12}$ levels that are more commonly caused due to malabsorption of protein-bound $B_{12}$ (food-bound $B_{12}$ malabsorption) rather than due to pernicious anemia. Food-bound $B_{12}$ malabsorption is found in people with certain gastric dysfunctions, particularly atrophic gastritis with low stomach acid secretion, which prevails among elderly people [2]. As the bioavailability of crystalline (free) $B_{12}$ is not altered in people with atrophic gastritis, the Institute of Medicine recommended that most of the RDA (2.4 μg/day) should be obtained by consuming foods fortified with $B_{12}$ or supplements containing $B_{12}$ [2].

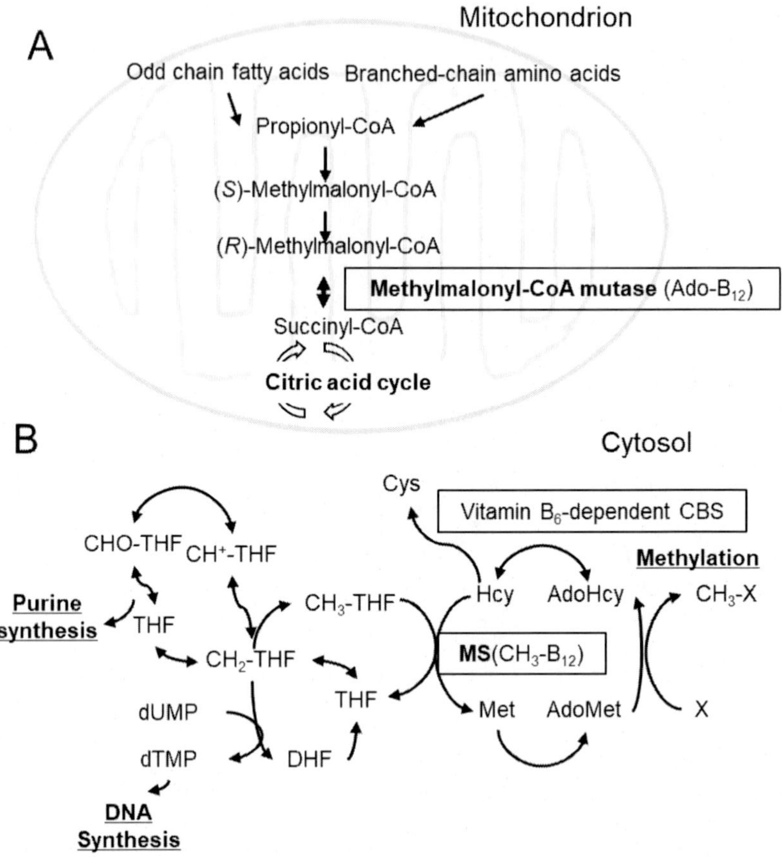


Figure 2. Vitamin $B_{12}$-dependent metabolic pathways in humans. (A) Catabolic pathway of odd-chain fatty acids and branched-chain amino acids. (B) Methionine synthetic pathway. Abbreviations: AdoHcy, *S*-adenosylhomocysteine; AdoMet, *S*-adenosylmethionine; CBS, cystathionine ß-synthase; CHO-THF, 10-formyl-tetrahydrofolate; $CH_2$-THF, methylenetetrahydrofolate; $CH_3$-THF, methyltetrahydrofolate; $CH^+$-THF, methenyltetrahydrofolate; Cys, cysteine; DHF, dihydrofolate; dTMP, deoxythymidylic acid; dUMP, deoxyuridylic acid; Hcy, homocysteine; Met, methionine; THF, tetrahydrofolate.

Thus, if a sufficient amount of free $B_{12}$ can be incorporated into vegetables, it would become an excellent source of $B_{12}$ for vegetarians and elderly people with atrophic gastritis. Therefore, if plant foods, primarily vegetables, can reinforce $B_{12}$, it can overcome the nutritional defect of plant foods. In this chapter, we describe the production and characterization of $B_{12}$-enriched lettuce leaves.

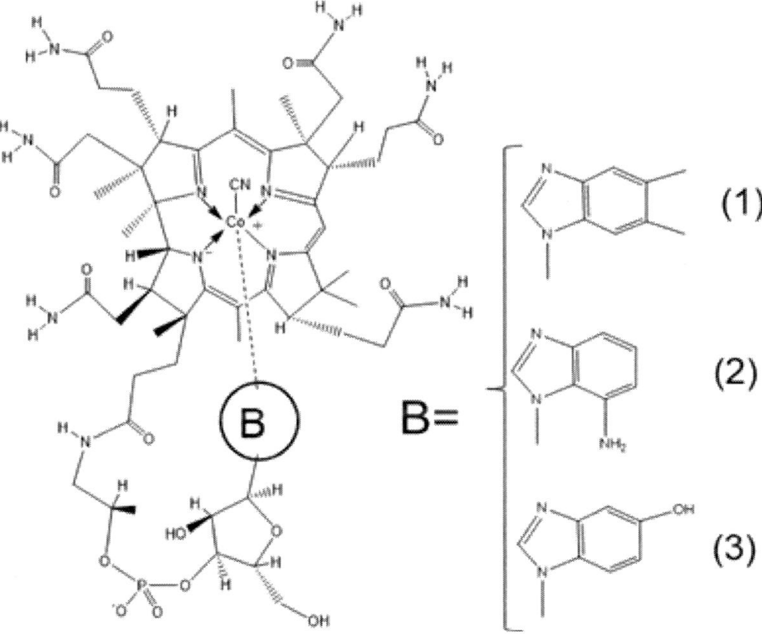


Figure 3. Structural formula of vitamin $B_{12}$ and partial structures of some inactive cobamides found in organic fertilizers and biofertilizers. (1) Vitamin $B_{12}$, (2) Pseudovitamin $B_{12}$, and (3) 5´-Hydroxybenzimidazolyl cyanocobamide.

## CORRINOIDS IN ORGANIC FERTILIZERS AND BIOFERTILIZERS

Mozafar reported that the $B_{12}$ content of spinach leaves (approximately 0.14 μg/100 g fresh weight) was significantly increased after using an organic fertilizer such as cow manure [10]. However, the majority of organic fertilizers, particularly those prepared from animal manures, contain considerable amounts of inactive corrinoid compounds in humans (our unpublished data) [11] (Figure 3). There are also reports describing the presence of such inactive corrinoid compounds in human feces, which account for >98% of the total corrinoid content [12].

Biofertilizers are products containing living cells of beneficial microorganisms, which can accelerate and improve plant growth by providing nutritionally important elements such as nitrogen and phosphorus. Nitrogen-fixing purple photosynthetic bacteria were reported to possess the ability to synthesize $B_{12}$ *de novo* [13]. We had previously analyzed the $B_{12}$ compound of three commercially available biofertilizers containing purple photosynthetic bacteria by liquid chromatography electrospray ionization mass spectrometry (LC/ESI-MS/MS) and identified Factor III (5-hydoxybenzeimidazolyl cobamide) and pseudovitamin $B_{12}$ (7-adenyl cobamide) as the major corrinoid compounds [11]. These results indicate that both organic fertilizers and biofertilizers are not suitable for use as $B_{12}$ sources to produce $B_{12}$-enriched lettuce leaves.

## PRODUCTION OF VITAMIN $B_{12}$-ENRICHED LETTUCE LEAVES USING HYDROPONIC SYSTEM

Hydroponic cultivation is an emerging technology that allows better control of water and nutrient supply, improves plant productivity, and reduces the use of pesticides. If a sufficient amount of free $B_{12}$ can be incorporated into lettuce leaves grown under hydroponic conditions, it would become an excellent source of free $B_{12}$ for vegetarians and elderly people.

### Production of $B_{12}$-Enriched Lettuce Leaves

As CN-$B_{12}$ is the most chemically stable form [1], it was used to incorporate $B_{12}$ into lettuce leaves. Sufficient quantity of $B_{12}$ was absorbed into lettuce leaves under hydroponic conditions [14]. Briefly, seeds of a butter-type head lettuce (*Lactuca sativa* L.) were germinated and grown for 24 days in black plastic pots containing sandy soil with an adequate supply of water. Then, uniform seedlings were selected and transferred into 3-L hydroponic lightproof pots (1 plant/pot) containing commercial nutrient solutions (Otsuka House series no. 1 and 2, Otsuka AgriTechno Co., Ltd.,

Tokyo, Japan). The nutrient solutions were renewed every 5 days. To incorporate $B_{12}$ into lettuce leaves, $B_{12}$ was added to the nutrient solution at various concentrations (1, 5, and 10 μM) 30 days after transplantation. Thereafter, the lettuce leaves were left for 24 h and then harvested.

## Characterization of Vitamin $B_{12}$ in Vitamin $B_{12}$-Enriched Lettuce Leaves

The $B_{12}$ content of the $B_{12}$-enriched lettuce leaves was determined using a microbiological method based on *Lactobacillus delbrueckii* sp. *lactis* ATCC 7830. Results showed that lettuce leaves treated with 1 μM $B_{12}$ contained 3.86 μg of $B_{12}$ per 100 g fresh weight [14], indicating that consuming approximately two stocks of these lettuce leaves can meet the RDA for adults (2.4 μg/day). Moreover, lettuce leaves treated with 5 and 10 μM $B_{12}$ contained approximately 16. 5 and 15. 5 μg of $B_{12}$ per 100 g fresh weight [14], respectively, which indicated that consuming two or three of these lettuce leaves would be sufficient to meet the RDA. The $B_{12}$ absorbed by the leaves was not converted into other $B_{12}$ compounds such as the coenzyme forms of $B_{12}$ or the degradation products of $B_{12}$ [14].

To apply this method in actual plant factories, $B_{12}$-enriched crisphead lettuce leaves were prepared using the circulation system of the hydroponics apparatus in a semi-large scale (40-L nutrition solution) (Figure 4). After the addition of $B_{12}$ to the nutrition solution at a concentration of 1 μM, the $B_{12}$ content of the treated lettuce leaves was approximately 3.6 μg/100 g fresh weight (our unpublished data). This value was similar to that of the lettuce leaves prepared in the above-described small-scale experiments. Since the $B_{12}$ content of the nutrition solution remained at >99% of the added $B_{12}$ content even after 6 times of continuous treatment of lettuce leaves, this $B_{12}$-containing nutrition solution could be used repeatedly.

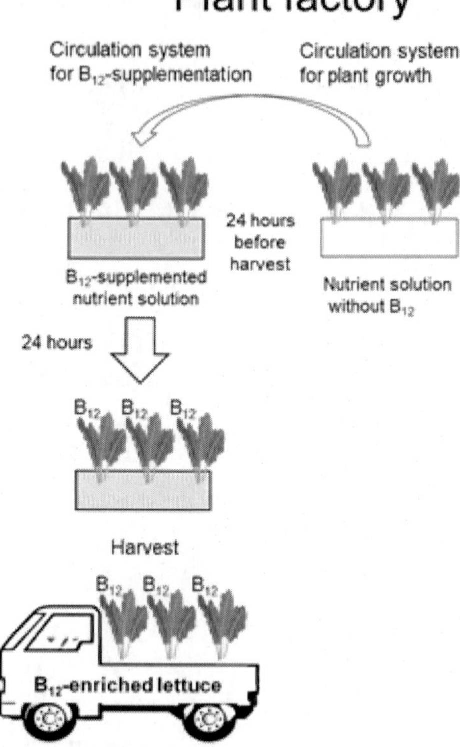


Figure 4. Preparation of vitamin $B_{12}$-enriched lettuce leaves using the circulation system of the hydroponic apparatus in plant factories. Seeds of a crisphead lettuce were germinated and grown for 24 days in black plastic pots containing sandy soil with an adequate supply of water. Uniform seedlings were then selected and transferred into the 40-L hydroponic circulation system containing commercial nutrient solutions. The nutrient solutions were renewed every 7 days during 30 days after transplantation. The plants were then transferred into the 40-L hydroponic circulation system containing 1 μM $B_{12}$-supplemented nutrient solutions and left for 24 h and then harvested.

## Effects of Vitamin $B_{12}$ Supplementation on Freshness of the Treated Lettuce Leaves

Lettuce leaves treated with 1 μM $B_{12}$-containing solution and then stored at15°C for 14 days showed a reduction in wilting and yellowing compared to that in control lettuce leaves (without $B_{12}$ treatment) [15]. The content of vitamin C, as an index of freshness of vegetables, were 11.47 ± 2.55 and 12.93 ± 2.78 mg/100 g fresh weight in the control and $B_{12}$-enriched lettuce leaves, respectively [15]. Although the difference between these vitamin C values was not statistically significant, the reduction of wilting and yellowing in the $B_{12}$-enriched lettuce leaves may be due to their slightly higher level of vitamin C. However, the mechanism by which the addition of $B_{12}$ increased the level of vitamin C level in the lettuce leaves grown under hydroponic conditions is not clear.

## OTHER VITAMINS OF LETTUCE

The various nutrients, including vitamins, present in lettuce (*L. sativa* L.) are absolutely required for various cellular metabolisms in humans [16]. Table 1 summarizes the reported vitamin contents of lettuce leaves. In particular, $B_{12}$, folate, and vitamin $B_6$ play important roles in homocysteine metabolism in human cells as shown in Figure 2B. As homocysteine possesses a pro-oxidant activity [17], deficiency of such vitamins can disrupt the cellular redox homeostasis to induce oxidative stress, which has been implicated in various human diseases, including atherosclerosis and neurodegenerative disorders [18]. The physiological function of $B_{12}$ is closely associated with folate and vitamin $B_6$, which acts as DNA and purine syntheses and as a cofactor of cystathionine ß-synthase (CBS), respectively. The contents of the antioxidant vitamins C and E are also shown in Table 1.

## CONCLUSION

$B_{12}$-enriched lettuce leaves were easily prepared in this study by treating with 5 μM $B_{12}$ for 24 h and growing them under hydroponic culture conditions for 30 days. Although we had attempted to incorporate $B_{12}$ into cherry tomato under the same hydroponic conditions, we failed to produce a $B_{12}$-enriched cherry tomato (unpublished data). Regarding the application of the hydroponic system, leafy vegetables such as lettuce are suitable for the preparation of $B_{12}$-enriched vegetables.

**Table 1. The contents of folate, vitamin C, vitamin E, and vitamin $B_6$ in various types of lettuce**

| Lettuce type | Total Folate | Vitamin C | Vitamin E | Vitamin $B_6$ | References |
|---|---|---|---|---|---|
| | (μg/100 g FW*) | (mg/100 g FW) | (μg/100 g FW) | (μg/100 g FW) | |
| *Crisphead* | 30-40 | - | - | - | [22] |
| | - | 4.2 | - | - | [23] |
| | - | - | 220 | - | [24] |
| | 30 | 2.8 | 180 | 42 | [25] |
| *Butterhead* | 70 | - | - | - | [22] |
| | - | 6.1-6.7 | 210-750 | - | [26] |
| | 70 | 3.7 | 180 | 82 | [25] |
| *Romaine* | 22 | - | - | - | [27] |
| | 60 | - | - | - | [22] |
| | 30-60 | 7-10.9 | - | - | [28] |
| | - | 2.8 | - | - | [23] |
| | - | - | 550 | - | [24] |
| | 140 | 4 | 130 | 74 | [25] |
| *Green Leaf* | 70 | - | - | - | [22] |
| | - | 6.5-30.2 | - | - | [29] |
| (Green oak) | - | 7.1 | 240-740 | - | [26] |
| | 40 | 9.2 | 220 | 90 | [25] |
| *Red Leaf* | 90-130 | | - | - | [22] |
| (Lollo rosso) | - | 11.7 | | - | [23] |
| (Red oak) | - | N. D.** | 240 | - | [26] |
| (Red oak) | - | 14.6 | - | - | [23] |
| | 40 | 3.7 | 150 | 100 | [25] |

* FW represents fresh weight.

** N. D. means not detected.

The recommendation of the Institute of Medicine that the majority of RDA for elderly people should be obtained from foods fortified with $B_{12}$ or supplements containing $B_{12}$ [2] was based on reports that approximately 30% of people aged >50 years are estimated to have food-bound $B_{12}$ malabsorption [2, 19, 20]. Therefore, $B_{12}$-enriched lettuce leaves could particularly become an excellent source of free $B_{12}$ supply for the elderly people, as the majority (86%) of $B_{12}$ found in the $B_{12}$-enriched lettuce leaves was free $B_{12}$ [15].

As shown in Table 1, lettuce leaves contain considerable amounts of folate and vitamin $B_6$. Vegetarian foods reportedly contain a high folate content [21]. Therefore, if a sufficient amount of $B_{12}$ is supplied to the human body via consumption of $B_{12}$-enriched lettuce leaves, serum homocysteine levels might be decreased in vegetarians due to the stimulation of $B_{12}$-dependent metabolic pathways as shown in Figure 2B.

## REFERENCES

[1] Watanabe, F., and Miyamoto, E. 2003. Hydrophilic Vitamins In: *Handbook of Thin-Layer Chromatography*, 3rd ed; Sherma, J., Fried, B., Eds. Marcel Dekker, Inc.: New York, NY, USA, 589-605.

[2] Institute of Medicine. 1998. Vitamin $B_{12}$. In: *Dietary Reference Intakes for Thiamin, Riboflavin, Niacin, Vitamin $B_6$, Folate, Vitamin $B_{12}$, Pantothenic Acid, Biotin, and Choline*; Institute of Medicine, National Academy Press: Washington, DC, 306-56.

[3] Fenton, W. A., Hack, A. M., Willard, H. F., Gertler, A., Rosenberg, L. E. 1982. Purification and properties of methylmalonyl coenzyme A mutase from human liver. *Archives of Biochemistry and Biophysics* 228:323-9.

[4] Chen, Z., Crippen, K., Gulati, S., Banerjee, R. 1994. Purification and kinetic mechanism of a mammalian methionine synthase from pig liver. *Journal of Biological Chemistry* 269:27193-7.

[5] Toyoshima, S., Watanabe, F., Saido, H., Miyatake, K., Nakano, Y. 1995. Methylmalonic acid inhibits respiration in rat liver mitochondria. *Journal of Nutrition* 125:2846-50.

[6] Watanabe, F., Yabuta, Y., Tanioka, Y., Bito, T. 2013. Biologically active vitamin $B_{12}$ compounds in foods for preventing deficiency among vegetarians and elderly subjects. *Journal of Agricultural and Food Chemistry* 61:6769-75.

[7] Watanabe, F. 2007. Vitamin $B_{12}$ sources and bioavailability. *Expermental Biology Medicine* 232:1266-74.

[8] Millet, P., Guilland, J. C., Fuchs, F., Klepping, J. 1989. Nutrient intake and vitamin status of healthy French vegetarians and nonvegetarians. *American Journal of Clinical Nutrition* 50:718-27.

[9] Pawlak, R., Parrott, S. J., Raj, S., Cullum-Dugan, D., Lucus, D. 2013. How prevalent is vitamin $B_{12}$ deficiency among vegetarians? *Nutrition Reviews* 71:110-7.

[10] Mozafar, A. 1994. Enrichment of some B-vitamins in plants with application of organic fertilizers. *Plant Soil* 167:305-11.

[11] Bito, T., Ohishi, N., Takenaka, S., Yabuta, Y., Miyamoto, E., Nishihara, E., Watanabe, F. 2012. Characterization of vitamin $B_{12}$ compounds in biofertilizers containing purple photosynthetic bacteria. *Trends in Chromatography* 7:23-8.

[12] Allen, R. H., Stabler, S. P. 2008. Identification and quantitation of cobalamin and cobalamin analogues in human feces. *American Journal of Clinical Nutrition* 87:1324-35.

[13] Zappa, S., Li, K., Bauer, C. E. 2010. The tetrapyrrole biosynthetic pathway and its regulation in *Rhodobacter capsulatus*. *Advances in Experimental Medicine and Biology* 675:229-50.

[14] Bito, T., Ohishi, N., Hatanaka, Y., Takenaka, S., Nishihara, E., Yabuta, Y., Watanabe, F. 2013. Production and characterization of cyanocobalamin-enriched lettuce (*Lactuca sativa* L.) grown using hydroponics. *Journal of Agricultural and Food Chemistry* 61:3852-8.

[15] Bito, T., Yabuta, Y., Nishihara, E., Watanabe, F. 2016. Effect of low temperature storage on freshness and vitamin $B_{12}$ content of harvested vitamin $B_{12}$-enriched lettuce (*Lactuca sativa* L.). *Vitamins* (Japan) 90:426-32.

[16] Kim, M. J., Moon, Y., Tou, J. C., Mou, B., Waterland, N. L. 2016. Nutritional value, bioactive compounds and health benefits of lettuce (*Lactuca sativa* L.). *Journal of Food Composition and Analysis* 49:19-34.

[17] Andrzej, J. O., Kilmer, S. M. 1993. Homocysteine metabolism and the oxidative modification of proteins and lipids. *Free Radical Biology and Medicine* 14: 683-93.

[18] Jessica, A. S., Nynke, E. H., Henk, J. B., Sinead, M. L., Coen, D. A. S., Jan, A. R., Paul, A. J. K., Victor, W. M. H., Hans, W. M. N. 2012. *S*-Adenosylhomocysteine induces apoptosis and phosphatidylserine exposure in endothelial cells independent of homocysteine. *Atherosclerosis* 221:48-54.

[19] van Asselt, D. Z., van den Broek, W. J., Lamers, C. B., Corstens, F. H., Hoefnagels, W. H. 1996. Free and protein-bound cobalamin absorption in healthy middle-aged and older subjects. *Journal of the American Geriatrics Society* 44:949-53.

[20] Krasinski, S. D., Russell, R. M., Samloff, I. M., Jacob, R. A., Dallal, G. E., McGrandy, R. B, Hartz, S. C. 1986. Fundic atrophic gastritis in an elderly population: effect on hemoglobin and several serum nutritional indicators. *Journal of the American Geriatrics Society* 34:800-6.

[21] Watanabe, F., Yabuta, Y., Bito, T., Teng, F. 2014. Vitamin $B_{12}$-containinng plant food sources for vegetarians. *Nutrients* 6:1861-73.

[22] Johanssen, M., Jagerstad, M., Frolich, W. 2007. Folates in lettuce: a pilot study. Scand. *Journal of Food and Nutrition* 51:22-30.

[23] Llorach, R., Martínez-Sánchez., Tomás-Barberán, F. A., Gill, M. I., Ferreres, F. 2008. Characterization of polyphenols and antioxidant properties of five lettuce varieties and escarole. *Food Chemistry* 108:1028-38.

[24] Chun, J., Lee, J., Ye, L., Exler, J., Eitenmiller, R. R. 2006. Tocopherol and tocotrienol contents of raw and processed fruits and vegetables in the United States diet. *Journal of Food Composition and Analysis* 19:196-204.

[25] USDA, 2015. *National Nutrient Database for Standard Reference Release 28.* USDA, Washington, D. C.

[26] Nicolle, C., Carnat, A., Fraisse, D., Lamaison, J. L., Rock, E., Michel, H., Amouroux, P., Remesy, C. 2004. Characterization and

variation of antioxidant micronutrients in lettuce (*Lactuca sativa* folium). *Journal of Science of Food and Chemistry* 84:2061-9.

[27] Wang, C., Riedl, K. M., Schwartz, S. J. 2013. Fate of folates during vegetable juice processing-deglutamylation and interconversion. *Food Research International* 53:440-8.

[28] López, A., Javier, G. A., Fenoll, J., Hellin, P., Flores, P. 2014. Chemical composition and antioxidant capacity of lettuce: Comparative study of regular-sized (Romaine) and baby-sized (Little Gem and Mini Romaine) types. *Journal of Food Composition and Analysis* 33:39-48.

[29] Koudela, M., Petríková, K. 2008. Nutrients content and yield in selected cultivars of leaf lettuce (*Lactuca sativa* L. var. *crispa*). *Horticultural Science* 35:99-106.

# INDEX

## A

## B



## C

## D

## E



## F



## G

## H



## I



## L

## M



## N



## O

## T



## U



## V

## W